Mark Time has made overcoming adversity an art form. The Royal Marines was his first metaphorical Goliath, but with fortitude and resilience as his sidekick he became a commando by the age of 17. Camaraderie, adventure and the occasional bout of idiocy have become personal watchwords creating his unique style that reflects the alternate face of society. In battling his own mental health issues, Mark is keen to provide humour in all his work, and is passionate about painting the world with colour.

In addition to books, Mark writes for a number of satirical websites, is a feature writer for the national press and a bumbling tech biff trying to keep a travel blog.

When not glued to his computer, Mark spends his time travelling, failing miserably to retain his six-pack and retrieving his hyperactive Jack Russell from rabbit burrows.

Mark grew up in Yorkshire but now divides his time between the UK and anywhere cheap.

Other Titles by Mark Time

GOING COMMANDO

www.marktimeauthor.com
facebook: Mark Time Author
twitter: @MarkTimeAuthor
instagram: marktimetravel

Praise for Going Commando:

'Tough, touching, it's a right good read' ~ The Daily Telegraph

'A cover to cover laughathon' ~ Soldier Magazine

'As exceptional as the Royal Marines themselves' ~ Weekend Sport

'A great book for Dads' ~ Huffington Post

'IMHO one of the best military books ever written. It's utterly hilarious' ~ Chris Thrall, Author

'Hilarious, yet truthful account of what it takes to be a Royal Marines Commando' ~ Rob Maylor, Author 'SAS Sniper'.

'You managed to reduce a quasi-misanthropic curmudgeon like me to a teary-eyed mess' ~ KP, Director International Governance and Risk Institute

GOING ALL THE WAY

PART II
OF THE BOOTNECK THREESOME TRILOGY

BY
MARK TIME

Published by Smashed Plate
Davington Mill, Bysing Wood Road
Favershan, Kent ME13 7UD

www.smashedplate.co.uk

ISBN: 978-0-9935470-0-3

Cover design by www.golden-rivet.com

Dedicated to those who paint with colour

Foreword

My final report complete, the working day was done. The *asr* call to prayer floated out from a dozen nearby minarets, permeating the scorched air as the shadows lengthened from the molten sun sinking behind the flat rooftops of Kuwait City.

I turned off my computer screen and turned on the TV. BBC World News flickered into life and I took in the headline: 'Marines bullying video condemned'.

There, shot on a poor quality mobile phone camera, were two naked Royal Marines or 'bootnecks' as they like to be called - their backsides pixelated in the name of good taste - fighting for the amusement of others. They flailed at each other desperately, with only forces-issue roll mats wrapped around their arms for protection, inside a muddy circle of other bootnecks who were openly laughing and cheering them on.

I was aghast.

Not at the naked fighting - that was just a typical day. What shocked me was the outcry, led by a brigade of journalists, MPs and moral snipers with absolutely no clue of how Royal Marines live their lives.

This was absolutely not, as was being reported, an 'initiation ceremony' to 'bully young recruits'. These young lads were fully-fledged commandos, old enough to die in combat but young enough to be refused a hire car, just back from taking part in a highly questionable war. They had seen things that would turn grey the hair of the most emotionally impervious stoic, and been ordered by the unaccountable ghosts of politics to commit acts that no human being should ever have to perpetrate. What was playing out on my TV screen - and the screens of a hundred million other viewers around the world - was an act of communal catharsis in the apparent privacy of their own barracks, away from the prying eyes of those who would not, and perhaps could not, comprehend.

Within a few days, those naked arses were in the papers and all over the internet. The chorus of disapproval was growing. Of course, the only people who publicly condemned such activities had never worn a green beret and this was an ideal opportunity to seize some airtime on issues they didn't understand but could attempt influence from afar. Those same MPs, who promoted anything to ensure their own survival not the welfare of our nation, watched the war with a whisky in their hand, ensconced within their big leather armchairs, pontificating their disgust while our lads returned with body parts missing; one commenting that two corporals in the ring of voyeurs, one dressed as a nurse with overly large balloon breasts, the other as a school girl, 'Weren't very dignified.' Neither, I declare Minister, is war.

Especially galling was the criticism by an ex British Army colonel who had been revered for his work in Bosnia; apparently forgetting he had ever been a soldier, claimed that the footage was 'shocking'. 'Shocking' could be apportioned to atrocities encountered in the Balkans conflict, but a few men skylarking about? Lauding himself as an 'expert', - evidently not in the field of perspective - he was clearly an outsider to the Royal Marines. If he knew anything about bootnecks he wouldn't have asked, 'Why are they naked for goodness sake?' How coincidental then, he was vilifying such activities just as he was about to start his own political career. And looking at him, he could never do the commando course.

Bootnecks, like all our teeth arm combatants, are constantly ready to face death. I think it's only reasonable they be allowed to fully celebrate life. They may shock through their antics, and they fight so that society has the right to comment. But Royal Marines have a right not to listen. Their comrades are their only legitimate critics.

So when I decided to write a series of books about the 'Corps', I did so with unashamed empathy borne through insight.

This is my journey through early adulthood, with the Royal Marines as my big brother, father, uncle, and, by the way some bootnecks dressed, - a pissy old aunt.

It is not a book about combat. Thankfully, I never witnessed the horrors of war. Peacekeeping was in vogue when I joined up. When I left some years later, the world was still quietly resting, having a breather, before we decided to yet again kick ten bells of shit out of each other.

During my relatively short career, I have been honoured to rub shoulders with men who have become TV personalities, captains of industry, doctors, and leading academics. I have shared beers with those who have conquered both poles, climbed Mount Everest, become renowned maritime explorers; coached, managed, and strengthened professional sports teams, international sports stars and won Olympic medals (although it was only a bronze - rubbish). This doesn't even touch upon those whose military exploits transcend the believable.

Despite what vox pop culture may think, the Royal Marines do not recruit ill-educated Neanderthals who only join up to escape the dole queue. In modern times,

40% of non-commissioned ranks are qualified to be officers. We have many marines - the lowest of all ranks - who have a Masters degree. We even have a former professor who once worked on the Hadron Collider. Be under no illusion, these are extraordinary men. Men, who make up a tiny, insignificant percentage of the country's population, yet make a huge, significant difference to its credibility - whether they occasionally fight naked in the mud or not.

I was never one of these superstars. I never looked upon myself as a hero. But as I look through spectacles of cynicism where before I looked through the rose tinted glasses of youthful exuberance, I now fully appreciate my time amongst these folk. Like the few who wear the green beret without making the headlines, I was a Royal Marines Commando, and for that, I will always be proud.

Author's Note

This is a personal reflection of my early years travelling the world as a young bootneck with testicles full of silliness, over a six year period when I managed to spend a whole 14 weeks at my UK base. I was paid around £60K in total through those years but please don't think I wasted it all. As I came out of this period with hardly a pot to piss in, I can honestly say it was all joyfully ploughed back into the economy through various establishments of pleasure, travel agencies and outdoor recreation businesses. Admittedly, much of my wages went overseas into the pockets of foreign business owners and for that I can only apologise on behalf of the taxpayer, but in fairness it was our politicians who insisted I travel to such horrible places as the Mediterranean, the Caribbean and the USA.

Moreover, it's a story to show that the folly of youth shouldn't define who we become. While stories may have been slightly embellished and 'dits' blended in to fit within the rhythm of the book, they are all based on fact and names, including my own, have been changed to protect the not-so-innocent. As in the first book of this series, *'Going Commando'*, it is written through the viewpoint of a naïve young man, and while I now may have slightly matured, I thought it important to be true to my young self, even though my opinions may have subsequently altered, especially where conflict, the objectification of women and pissing in your own face are concerned.

As if naked roll mat fighting was the worst thing a bootneck has ever done…

Mark Time

Acknowledgements

My heartfelt thanks go to those who have assisted in my development as an author since the release of 'Going Commando'. Success would never have been possible without your guidance.

Once more, indescribable gratitude is given to the creative genius that is Andy Screen of 'Golden Rivet', your benevolence of spirit is immeasurable.

To Chris Thrall, author of the best selling memoir 'Eating Smoke', who dragged me from the literary pyre to continue my passion.

To those closest to me, your friendship, encouragement and empathy during my darkest days will never be forgotten. You know who you are.

Finally, to Jo, Connor and Finlay without whose unwavering support I may not have lived to write as I do.

*'Be wicked, be brave, be drunk, be reckless, be dissolute,
be despotic, be a suffragette, be anything you like,
but for pity's sake be it to the top of your bent.
Live fully, live passionately, live disastrously.'*

~ Violet Trefusis, writer

'Royal Marines should be exiled to a desert island with only loose women and alcohol for company, only allowed off in times of national emergency.'

~ *Mary Whitehouse, social activist*

THOSE FAMILIAR butterflies tickled my stomach just as they did on my first day of training, when I alighted the train platform welcomed by the sunshine of Taunton - home of 40 Commando Royal Marines.

A picturesque country town and administrative centre for the county of Somerset, Taunton has an air of an overgrown country village; a once thriving market square, resplendent in council concrete and blooming borders contrast the odd remaining castle keep that ensnare history buffs into its languid bosom.

40 Commando RM had been in Taunton since 1983, or stapled onto the small outlying village of Norton Fitzwarren, to be exact. As the name suggests, Norton Manor Camp was once owned by the local gentry; its commanding manor house now the officers' mess standing high above the rest of the camp. Its previous role was one of being a Junior Leaders camp for the British Army's 16 and 17 year olds training for their careers as adults. To the local thugs these young lads were easy targets, and many young squaddie had been beaten up as a result. When 40 Commando RM arrived, the unit was on a ship bound for exercises around the Mediterranean Sea, so a small rear party was given the task to move the remaining stores, equipment and paperwork up from Seaton Barracks in Plymouth to their new Somerset home. As a rear party they were few in number, and the local ruffians thought that 40 commandos were now the new permanent camp

incumbents. 40 commandos, no matter how tough, would not be able to stand up to a whole town's worth of local hard men. Often the odd bootneck visiting town was given a bit of a going over, the thugs confident that the other 39 wouldn't seek revenge. Unbeknown to them 40 Commando RM main party had now returned from the Med, and the thugs had seriously miscalculated, not realising that 650 commandos had now arrived. When news of these indiscriminate attacks reached the returning members, swathes of marines entered town and meted out swift retribution. Rather alternative to a honourable welcome march, I suppose.

On entering Norton Manor Camp, I thought I'd gone back to the 1950's. The main gate sentry dressed in full lovat uniform stood smartly in front of creosote smelling wooden huts; the guardroom designed by Churchill's dad. Accompanied by Jay, a fellow Kings Squaddie with whom I'd passed out of Royal Marines basic training, the Guard Commander warmly greeted us, checking our paperwork to confirm we weren't lost and trying to find the local youth club.

A marine walked towards us, his green beret in his overly-ironed green denim's pocket. From within the confines of the guardroom out strode a brute of a man with an even louder voice. In his hand he held a thin cane tipped with a nickel thimble cap signifying him as the Provost Sergeant.

Thankfully, his booming words weren't directed at us. 'Oi lofty, where's your beret?'

'It's here in my pocket, Colours.' The lad, although clearly a commando, didn't look much older than me.

'Why is it not on your head?'

'I got a no headdress chit, Colours. I got my head cut open playing rugby.'

'Ah right, hold on there then.' The Provost Sergeant returned after a short trip back inside the guardroom. 'Let me have a look at your chit.'

The young marine took out from his pocket a small piece of paper written by the camp's Medical Officer excusing the wearing of headdress.

The Provost Sergeant examined it carefully. 'OK, come here.'

The young marine closed in on the Provost Sergeant who, in one swift movement that belied his size, reached into his pocket, pushed the chit to just below the young man's hairline and stapled it to the marine's forehead.

'OK, you can wear your chit instead. Now, what do you want?'

Jay and I exchanged glances. Was this how life in a unit was to be? We'd heard life was pretty relaxed, the fruit we could now eat after the labours of our training.

The Guard Commander smiled, pointing us in the direction of the transit accommodation. 'Welcome to 40.'

My nostrils were aroused with early summer's amalgam of desiccated aromas: pollen, dust, sundried wood treatments and forgotten carbolic soap. I walked along the threadbare carpet laid over creaking floorboards and wondered whether this was the dream I'd envisaged. The wooden transit accommodation was sparse, with pre-prepared bedding similar to that I'd been issued at Commando Training Centre for those many elongated months. With the whole evening spare before reporting to the Movements Sergeant the next morning, we sat on our beds pondering what we should do. All options, we decided, were too dangerous. Despite being newly qualified commandos there was nothing more frightening than being in a camp of 650 other commandos. As a result, we dared not venture out any further than the heads, and only then once we'd done a 'recce' to ensure there wasn't any other green bereted loons having a piss.

Late in the evening, an RAF corporal joined us. He was spending a fortnight here before moving to more permanent accommodation at a nearby RAF communications base. We said little to him, not knowing whether to call him 'Corporal' as we were just out of training or 'mate' as we were fully trained dealers of death who laughed in the face of danger (as long as it wasn't the face of a fellow bootneck). In fairness, he had little to say to us, other than request the direction of the NAAFI, somewhere we had considered visiting but thought we may get bummed on the way.

We rose early the next morning and with stomachs hungry from missing the previous night's dinner, decided to go to breakfast. The RAF corporal followed us into the heads wearing his uniform. Jay and I instantly gained height in stature watching him shave with his jumper on.

'We don't shave with our tops on,' I explained, echoing the words of my Drill Instructor while in the induction phase of basic training.

'I don't give a fuck mate. I'm not a Royal. Who are you again?'

'I'm Mark,' I said a little more indignantly than warranted. 'If you do that in the regular grots you'll end up with a regimental bath, one that consists of bleach, washing powder and a hard bristly bru…'

'…I know what a 'regi' bath is son,' interjected the corporal. 'Unlike you, I've been in longer than a NAAFI break. When I want advice from you I'll ask for it. Got it?'

Well and truly put in my place, I didn't really want to start my new career getting charged for arguing with an RAF corporal so I left the debate rather deflated.

I saw him a couple of days later. He'd moved temporarily to HQ company accommodation and had somehow lost both his eyebrows.

Welcome to 40 indeed.

***

40 Commando RM was my reward for passing 30 weeks of Royal Marines training. It'd taken me longer due, in part, to me being pretty shit at the many things required to be a commando. But overcoming adversity was one trait that the Royal Marines had taught me and, as such, after 36 weeks I'd eventually vanquished the foe of training becoming a green bereted commando at the age of 17 - a true 'bit of skin'. I was now a massive 5'6" and weighed a whopping 10st 3lbs. A poster boy for the Royal Marines I was not; more likely a candidate for modelling children's 'days of the week' Y-fronts for Freemans catalogues.

40 Commando was divided into Alpha, Bravo, and Charlie fighting companies; Support Company made up of heavy armament troops, and HQ Company, made up of additional operational, administrative sub units and those who wore spectacles. As young marines fresh from 'the box' Jay and I would be placed in a fighting company, just in case we did go to war and they needed disposable cattle fodder.

We reported to the Movements Sergeant who instructed that Bravo Company would be our new home. They'd just returned from a long Northern European exercise, and due to utter boredom on the civilian STUFT ship (Ship Taken Up From Trade) bobbing off the coast of Southern Norway had decided to lighten the mood. One marine requested over a number of announcements, that the sexily voiced Norwegian purser ask over the ship's tannoy for 'Marine P. Niss', 'Corporal Danny Twyer', and 'Sergeant Jerry Can' and the Dutch Marine on exchange 'Captain Naafi Van Driver'. The Regimental Sergeant Major (RSM), a little fed up with these recurrent joke announcements, warned the anonymous marine from Bravo Company to stop forthwith. After a pregnant pause, the next announcement from the innocent sexy purser requested, 'All M.U.F Divers come to Mike Hunt, all M.U.F divers come to Mike Hunt.'

Jay and I reported to Bravo Company Sergeant Major (CSM) as instructed. More nervous than I'd ever been as a recruit, I reported like a recruit does: springing to attention reciting number name and permission to speak. The Company Clerk, sat with the CSM, giggled as he'd probably done the many times new lads had entered. The CSM told us there was no need for such formalities now and I was welcomed to the company.

Unfortunately, I was in Bravo Company for possibly the shortest time ever. The CSM needed two volunteers to go to the officers' mess to be MOAs (Marine Officer's Assistants) - flunkies to nursemaid the officers. As it was unlikely he'd get willing volunteers for such a shitty job, our entrance into his office was timely. Sensing our disappointment, he assured us our time for being steely-eyed killers would come and he'd ensure our return to Bravo just before summer leave in eight weeks time. So, before I'd even done a joining routine for Bravo Company, I was joining HQ Company as a disgruntled bed maker and waiter. This was not what I had in mind.

The HQ accommodation block was situated next to the galley - convenient for meals, inconvenient should you not want the permanent smell of over-cooked onions and rotting slop. Like the other blocks it stood as a single storey wooden structure consisting of one central spine containing ablutions, drying rooms and ancillary spaces. Down each side of this central spine were three legs of accommodation. Because of their configuration these blocks were called 'spiders', a name dreamt up by somebody who'd clearly never counted the legs on a spider. The benefit of living in these decrepit WWII era blocks was that we paid very little for the privilege of living there, meaning more disposable income could be wasted. Each room was nicknamed a 'grot' - a man cave where the ten or so inhabitants would venture out only to seek pleasure and the occasional bit of work.

My first grot I shared with some drivers. Most were older marines, and the first man Jay and I met was a 6'5" monster built like two brick shithouses welded together. I recognised him from the Corps Boxing Championships the previous year. A fearsome looking creature, 'Tiny' was a true gentleman and cordially greeted us both. The others in the room were a mixture of drunks, deviants and desperados, clearly men for whom the Royal Marines was a labour of excesses according to their waistlines. As marines of many years, they'd passed the phase of initiations and welcoming committees preferring to secure their inactivity by using any new joiner as an automatic runner to the NAAFI or permanent wet (tea or coffee) maker.

I noted one of the lads in the grot wore an eye patch. Having suffered a similar injury during training, in a moment of bravado, I asked, 'Did you run into a tree as well?'

Confused, until I offered an explanation to my slightly weird introductory sentence, he said, 'No I have gonorrhoea.' And with that explanation opened up his patch to show a reddened eye, swimming in pus.

Admittedly, I wasn't particularly experienced sexually but pretty sure the eye wasn't part of the reproductive system. I then thought of the karma sutra and what position he'd have taken to get such an affliction. He was rather tall and skinny so accepted he maybe could bend into positions only degenerate gymnasts could attain.

'I got it from a bar girl in Thailand on Easter leave,' he explained. 'Last girl I shagged, I reckon, as well. Talk about bad luck. It's a lot better now.' His harsh Geordie accent made it sound far more comical than necessary. 'You been to Thailand?'

'Err, I have been to Flamingo land.' Well it did sound exotic when my uncle took me there when I was nine.

'And I doubt you've done any whores,' he said with contemptuous disdain, intonation probably more suited if he'd deleted the word 'whores' and inserted 'charity work'.

'None, I'm afraid.' And at my age, I would have been. The nearest I'd come to a prostitute was a girl at school who let boys finger her for a packet of *Quavers*.

Yet it was quite prophetic that within a week I would have my first taste (not literally) of a sexually transmitted disease.

Being new and extremely young, my first guard duty came pretty quickly. Patrolling the camp in pairs was immensely laborious yet I was keen to make sure I wasn't one of those who would sit in the warmth of the launderette or go to the grots to make tea. So when I did have the chance to sleep I did so with genuine tiredness. Despite Royal Marines being impeccably clean, the constant sharing of mattresses and bedding in the guardroom's off duty room meant that certain ailments could be transmitted through indirect contact. One of those ailments, it transpired, was pubic lice, commonly known as 'crabs'.

I'd completed my first duty on the Friday night. The main advantage of living with the drivers' was being next to the chefs' grot who had permanent hot water from the stolen, or should I say relocated, boiler, unlimited toast from the relocated toaster, and unlimited food plainly stolen from the galley. I'd managed four hours kip after guard so rose from bed at a similar time to the lads crawling out of theirs with a half eaten kebab as their lover. As usual, my first task, even after coming off guard duty, was to go and make the wets.

On entering the chefs' grot for some milk, there, on their issued sofa that displayed more stains than a church window, sat three chefs watching the start of Football Focus. Nothing out of the ordinary there, and even their nakedness was of no surprise. What did strike me as odd was a woman, equally bereft of clothing, on her knees fellating the lad in the middle. What made it more comical was that neither lad either side of the impassive chef was taking any interest in the sex act going on, but more interested in what Bob Wilson had to say. Even the lad in the middle was offering more advice to Kenny Daglish on the screen than the girl between his legs busily working her respectability into the gutter. I made the wets with the odd glance at the free peep show occurring. I have to admit, it was the most arousing thing I'd ever witnessed. At this point of self-analysis, I decided that I really needed to lose my virginity.

With the amount of girls invited back to camp it shouldn't have been a real issue, but I was 17 and the girls that did venture through the gate were far happier with men who knew what they were doing. If they were going to breach camp security a good rogering was the least they deserved. Although strictly illegal, it was often known for lads to bring girls onto camp should the female not want to go home for unspecified reasons, usually the presence of a husband in her bed. The boot of a car was the usual place of secretion, pointless really when the driver would just say, 'Don't bother checking the boot, there's a girl in there,' gaining access should they personally know the Guard Commander.

Women had been known to stay on camp for weeks on end, watching daytime TV in the accommodation surviving on semen and the meals brought back by the lad with whom she had a 'relationship'. Many would have to hide in a locked locker during Friday morning room inspections to retain their secrecy from the hierarchy, only to be released after the rounds had been completed, if the lads remembered.

After making the wets and taking another quick glance at the girl who seemed to be able to suck a golf ball through a hosepipe, I walked to the ablutions with a towel draped tightly around my midriff, trying not to the think of the girl's head bobbing between the hairy knees of Grandstand-loving chefs. As usual, the ablutions had that typical male bathroom smell, the one of shit and toothpaste. I took a second look at an arse that wasn't pert enough to be a bootneck's, to find yet another naked woman, this time having a shower.

Disgusting, I thought, *her showering at cost to the taxpayer.*

With the thoughts of blow job girl and showering girl in my head I had to release my pent up sexual fervour somehow, so Pamela Hand-erson and her five fingers joined me for some self-loving in the toilet cubicle - an area hardly runner up to Paris as a haven of romance. Unfortunately, there was other guest seemingly invited to my genitals. I saw movement, and it wasn't my glans waving about. Within the confines of my pubes, I saw the creeping about of something small; again it wasn't my bell-end. I plucked out and studied what I could only think of as a mite. Upon being squeezed it popped like a blister, so searched for more. It seemed the visitor was a lonely character so considered no further action and having wildlife in your pubes should be no reason to not finish off a good old wank.

That's the thing with crabs, it only takes one to start laying eggs and before you know it you are scratching and pulling like a demented chimp. That weekend I found more eggs glued to my pubes. Panic stricken, I did what any clueless moron would do - I shaved my pubes. Shaving is OK should you shave everywhere. My bum hole, as anyone who has ever seen it would attest, looks like an Alsatian's back, so while from the front I looked like a pre-pubescent schoolboy, my itching continued in my undercarriage and more worrying, I found one of the more adventurous little blighters on my upper thigh.

The upside of this affliction was cover for my virginal status. There were probably more necrophiliacs in 40 Commando than virgins, so I took every opportunity to show off my dirty ball sac. The lads in the grot were rather apathetic to the whole affair. They had seen more 'naughty diseases' than a pox doctor's clerk and offered me advice such as to boil wash every item of clothing and bedding, and not to squeeze crabs while sat on the settee watching the news which, under the circumstances, was a fair request.

Every week 40 Commando sickbay would offer a trip to the local hospital for Taunton's genito-urinary department, or as we called it - the 'VD clinic' (while now obsolete, I still think 'venereal' is a far more friendly word than the more clinical phrase 'sexually transmitted'. After all, the friendliness part was what usually led to us being diseased in the first place). I decided to report with what seemed a whole troop worth of marines - a farrago of uninhibited deviants with variable gaits depending on the severity of their affliction. Should I have not been in such a large number I may have felt shameful attending such a clinic, yet members like Danny the Geordie with his pus filled eye made the experience far less intimidating and rather communal, like an organised Rotary Club day out. We even got a free bus.

A visit to the genito-urinary department should never feel like a homecoming, yet the receptionist knew a few of the lads and exchanged an amiable chat with Danny upon his entering. I looked around and noted two civilians sat in the waiting room both hidden behind random magazines held far too close for comfortable reading. I approached the desk.

'Hello there, what can I do for you?' It was such a loaded question between her pearly white teeth that only the truth would be a sufficient answer.

'I think I have crabs,' I offered meekly.

'No problem,' she said, as I if I had thanked her for something. 'Name?'

'Mark Time.'

I could feel the magazine-eyed civilians look at me, the lads laughing behind.

'It's your first time here isn't it. We only need a first name. You know, for your own privacy.' Her smile bore into my embarrassment as I sat next to a bootneck who called me 'Mark Time - the most famous deviant in Somerset.'

In any examination room of a genito-urinary department there are a lot of academic pictures of cocks and fannies. Thankfully not quite as erotic as a picture of a 'Reader's Wife', I glared intently at one to take my attention away from the doctor who, despite my protestations of me retaining my virginity, proposed a full urethra scrape was necessary to check for any further diseases. Why are all genito-urinary doctors female? And to add more insult to injury, she was hot.

Known as an 'umbrella', but scientifically known as a urethral dilator, the instrument used to take cell samples from inside the todger, thankfully looked nothing like a parasol for a small rodent. In reality, it was probably only a couple of millimetres in diameter; however, as it neared my flaccid embarrassed manhood it took on the shape and size of a golf brolly. Within the smooth, latex hand of 'Miss Horny Doctor 1987', I feared an erection could be on the cards but anger was the last thing on its mind. Instead, it cowered in fear within the confines of an increasingly wrinkled foreskin.

I took my eyes away from the poster of a colourful cervix to glimpse at the instrument's insertion. The pain was tolerable despite it edging further up than I thought necessary. The extraction, however, wasn't so. Twisting the apparatus opened a small umbrella shape to collect the cells. It felt as if they were scraping away the soul of my nether regions and taking with it lumps of flesh that could be used for military rations. With a look on my face that could only be read as, 'For the love of God will you hurry up?' I winced as she withdrew the umbrella, placing the invisible sample into a receptacle labeled: 'Mark - apparent virgin', that would now be examined by probably more females who could at least say, 'Well at least it's not gonorrhoea of the eye.'

***

While gutted I was serving salads and Pimms to the commissioned ranks and on my days off visiting the 'pox doctor', fate dealt me a good hand by allowing me to gently settle into unit life in a job with little stress. The officers were surprisingly pleasant and respectful, the only bollocking ever offered was due to me incorrectly applying lemon to a gin and tonic; a lecture I'd obviously missed while undertaking commando training. My companions were rapscallions from other companies sent there as a punishment, so their lack of eagerness to help the officers was understandable. One of the lads, Hodgey, had nicked a double decker bus to get lads back to camp because of the drought of local taxis, so had a few months as an MOA to think about his choice of joyriding vehicles. As he'd been 'stitched up' - in his words - by his Troop Officer, he felt it only right and proper to gain a modicum of revenge by not only regularly inserting the Troop Officer's toothbrush up his anus, but also ordering a selection of hardcore Dutch pornography and sex aids to the address of the Troop Officer's mum. I; however, thought it best to refrain from sticking implements anywhere near my crab ridden genitalia but get my head down and crack on until summer leave arrived.

'We are mature in one realm, childish in another.'

~ Anaïs Nin, Novelist

TAUNTON HAS A RELATIVELY quiet town centre. Saturdays typically the busiest, a high street walled with recognisable shop fronts paraded for the passing shoppers swaddled in the uniformity of weekend consumerism reflecting the urban landscape nation-wide. What is less familiar at around 2pm is the sight of two sombrero-hatted men, naked other than a card-board box suspended by trouser braces, browsing Debenhams' window. This was my first glimpse of the Royal Marines tradition of 'silly rig' - fancy dress for commandos.

You can tell a lot about a Royal Marine from his locker. It's the chipboard gateway to a marine's con-science. Venture inside and you'll see how he leads his life. There are many who have smartly ordered clothing, both military and civilian, hanging in demarcated columns above equipment kept spotlessly clean. His shelves will be stocked with neatly folded T-shirts and underwear; sports kit allocated its own shelf as it is so often used. The whole locker smells of aftershave, shaving foam and toothpaste - a man who is prepared, knows what he wants in life and thinks ahead to get there. Then you have bootnecks who see life in the Royal Marines as one long party. Sure, there remains military uniforms and equipment, some issued but some civilian bought, the issued stuff left unused in favour of the better quality civilian equivalent. Smart civilian clothing will be either hung up for the next night out or crumpled in a corner smelling of beer and cigarettes from the previous one. The shelving predominantly stacks pornography complete with an old pusser's sock that is so stiff it feels as if it has a backbone (not that you would ever feel it, you know what it's covered in). The most deviant would stock a jar of Vaseline with the tell tale indentations of brown fingerprints, cigarettes and the odd bottle of spirits. Laid next to his washing gear is last night's half eaten takeaway or a pair of girl's knickers - a trophy yet to be placed on the gronk's board or to be worn, as they 'feel nice'.

Whether keen or not, one thing that is ubiquitous in a Royal Marine's locker, as sure as there will be a green beret, you will uncover the stereotypical clothing of prostitutes, train spotters, ballet dancers, super heroes and the like. 'Silly rig runs' - going out in fancy dress, is as much engrained into corps culture, as it is to celebrate the Royal Marines' birthday on the 28th October.

In my first few weeks as a trained marine, a whole company of bootnecks from 42 Commando dressed as Vikings, armed with plastic weaponry attacked 'Monroes' nightclub on Plymouth's Union Street in revenge for the bouncers beating up a couple of lads. What damage a plastic axe can do against a hymen of bouncers is anyone's guess, but I can only imagine it being hilarious. The sick bastards of 6 Troop Bravo Company, would gain notoriety and a million extra duties for dressing up in combats complete with toy AK 47s on a 'Michael Ryan run ashore' only weeks after the Hungerford massacre.

My early months in Bravo Company not only showed me advanced commando soldiering skills, but how to get the best deals in charity shops and even more importantly taught me the noble art of transvestitism.

My first real venture into silly rig was rather tame. While still in HQ Company, I'd made good friends with Jock who'd passed out a couple of weeks after me. He'd also been given a shitty job as a photocopier gofer, a job even worse than mine. Jock and I took the train down to Newquay to see Rick and Tug, two of my old civilian mates who were working there for the summer.

There is no finer place on Earth than a sunny England. Wan skin is presented to the sun as if being offered as sacrifice, ice cream stalls add colour through their rogues gallery of frozen delights to the promenades on which they stand, greyed landscapes are injected with a brilliant colourwash, and street swearing trebles in decibels from sun-worshipping outdoor drinkers whose redness deepens with each pint. It was on such a glorious day Jock and I boarded the train to Newquay. Beer and sun make a heady mix, even on public transport. As we'd decided to make the journey dressed as babies, by the time we'd left Par Station we'd regressed into infantile romping, crawling and burbling down the train aisles much to the disgust of those wanting a quiet weekend away. Towels wrapped as nappies did their jobs, but to sober noses the stench of urine and admittedly a bit of poo from drunken mis-wiping, would have put them further off their British Rail BLTs.

We alighted at Newquay to be greeted by our two compatriots, slightly taken aback by the sight that now wobbled in front of them. Like a baby, I'd been sick down my front, proving that baby bibs are meant for baby-sized stomachs; only I didn't wear a compote of pureed chicken in a carrot and basil dressing hanging around my neck like a Mr T necklace - mine was a compote of pureed beer and steak and kidney pie. Clearly I wouldn't be allowed into Newquay's finest public houses like this. I was far more suited to falling asleep at 6pm on my mate's kitchenette floor resplendent in baby clothing and bodily fluids.

If being chilled to the bone on Portland Bill five months earlier while attacking HMS Osprey as a recruit was the coldest I'd ever been, then swimming and pretending to be a Bondi lifeguard still wearing nothing but my stretched toweling baby hat and ridiculously gaudy beach shorts in the Newquay surf resulted in the worst sunburn I'd ever sustained. I was a bootneck, bulletproof but seemingly not sun proof, and sat chilled and shivering on the return train journey, while radiating such heat that those nearby took off their sweaters. Dehydrated and still nauseous, I started work the next morning wearing a blister that spanned the width of my shoulders that wept like a grieving mother every time I stretched. Stuck to my shirt by blister fluid I had to undress in the shower to prevent further tearing of the skin. The subsequent scars still remain. We learn from our mistakes and I was learning a lot in my youth.

***

With my two months penance in the officers' mess complete I was eagerly thrown back into 4 Troop Bravo Company. 4 Troop was a bit of an exaggeration, a troop should be made up of three eight-man sections plus a HQ element. 4 Troop totalled thirteen men. With this number at least I'd get a bed.

My chaperone to guide me around Bravo Company was a lance corporal called 'Big Gav' - a 'brown wings drifter' who saw anal sex as foreplay. He was yet another behemoth of a human specimen who looked as though he could strangle a bear. With his reputation preceding him one would think he should probably shag it first. He was truly a gopping creature with a penchant for waking up young marines with his large erection. 'Big Gav' was one of those wretches in the Corps everyone warned you about. His party piece was to drink a pint of urine with a turd dropped in. A special talent, he'd shown it the previous year at the Royal Tournament at the marines' bar. Amongst the audience were wildly impressed marines - and his mum who'd just popped in unannounced to say hello.

Big Gav showed me my bed space, my own private estate. I squeezed myself into the remaining space available, which just about accommodated a single bed, a locker and a flag to partition myself from the rest. Moving the locker to fit my bed uncovered some slight faults in the accommodation. Unless there were ravenous termites in Somerset, some previous demolition was evident. The wood panelling of the wall had somehow been removed, leaving a hole big enough for a man on a diet of Cumberland sausage to fit through. From where I lay, I could see other accommodation across a nice patch of grass. The upshot was that even without a window I had a nice view accompanied by the incessant cooling breeze that I presumed would become a tad disagreeable come November.

When the rest of the lads returned from physical training known as 'phys', I was greeted like a dose of herpes. I couldn't blame Steve, I'd taken up his rather large parcel of real estate and my rearranging of the spare locker had uncovered the large hole. Despite the rooms being designed for ten people Steve, like most lads, had taken up three people's worth and had opportunistically occupied a copious amount of room around their multiple locker area. They liked it that way - it was their sign of seniority. I was a sprog so thanked my lucky stars that my bed actually had springs, unlike Jay who slept bowed as if in a hammock, and would each morning hobble from his bed like a hunchbacked septuagenarian.

I'd thought life with the drivers was quite surreal, but had put that down to being new. Here I was, two months out of training, thinking I'd be more accustomed to the shenanigans of bootnecks. I was wrong.

My first morning in 4 Troop, I awoke early, not quite with excitement, but without alcohol, unlike everyone else. I waded through the rejectamenta of the previous night's frivolities - moist knickers (without females inside them), crumpled silver tins of half-consumed takeaways, diarrhoea spots, and the odd comatosed body sleeping rough on the comfort of inebriation. As per normal, I headed straight to the ablutions wearing only my wrap around towel. En route, I passed a large man, in both physique and penis, wearing a size 10 sheer pink baby doll nighty that did nothing to flatter; it certainly didn't colour match the smeared lipstick that spread across his cheeks.

'Morning,' he muttered nonchalantly, scratching his ball sac that protruded from the bottom of his lingerie.

'Err morning,' I replied, as if it was the most natural thing in the world.

Even wearing such lingerie he continued Royal Marine tradition by rolling the straps from his shoulders letting it fall to the floor while he stood at the sink. There was no way he was going to shave with a nighty on.

Shaving at the sink, the lad with lingerie around his ankles talked casually to his mate sat naked on the toilet. Having his morning constitution while chatting, breathing, blinking and eating the left overs of the previous night's curry, proved that males can multi task, certainly at the most primordial level.

After showering I returned and bumped into Clint - a man so named as he shared the surname with a certain film star.

'Morning.' These lads were all terribly polite.

'Morning,' I replied. 'Why are you carrying a mattress?' I didn't think it too untoward to ask.

'Sprog are we?' he said knowingly.

'Yeah, just come over from HQ Company,' I said.

'Well you'll get used to it here young Royal. It's basic swamp rat admin. I piss the bed as soon as I see a pint, so I have two mattresses. This one's covered in piss so I'm off to the drying room to swap it with one that has dried out.'

'Ah right.' *That's logical* I thought as I passed a blood-soaked bootneck picking glass from his forehead.

It was a strange start to a Wednesday morning.

A typical day in a fighting company was far removed from those unrelenting days at CTC where fear drove one to run around like a headless chicken from 'arsehole 'til breakfast' in various forms of ultra smart dress.

Here, in my wooden hut, I'd get up at around 0730. Shit, shave and shower.

Refreshed as one could be with remnants of curry still bubbling away in the large intestine, we'd saunter casually to the galley for a leisurely breakfast should there be nothing left over in the grots from a night out.

We would parade at 0815, usually in Phys rig to go for a long run, or undertake some circuits in the gym or on the football pitch, ignoring the condemned assault course that was only utilised in the dark when inebriated. Many of the lads would use this morning session of phys to expunge the alcohol from their traumatised bodies. Some would have drunk their own body weight in Drambuie the previous night yet here they were, under a cloud of flammable alcohol fumes, running at impressive speeds only stopping to vomit bile into the undergrowth of Somerset's country lanes. Shower.

A stand easy at 1000 would see us all either sat in the NAAFI eating gargantuan rolls and drinking pints of milk, or in the grots sitting around discussing ephemeral subjects such as who would win a fight between a bear with a scythe or a pterodactyl with nunchuks.

At 1045 at least one of the troop corporals would walk off holding a clipboard to try and look busy, pointing occasionally at nothing in particular. Us marines would parade again, this time for a lecture, usually taken by one of the candidates - marines looking for promotion. These lectures could see us reinforcing our knowledge on global affairs or confirmatory practice periods of weapon training or tactics. Should we even get to a level of exertion where a bead of sweat was just about to leave the epidermis we would then shower.

Lunch was from 1230-1400 or usually 1405 as 'Neighbours' would finish at 1400, and nothing other than an all out nuclear attack was going to prise any of us from our chairs when Kylie was on the telly.

At 1405 we would parade yet again for another lecture, unless quite senior then you could make an excuse of 'doing admin' - a rather un-secret code for going back to the grots and having a wet. This single period lecture would allow family men to get home just in time to pick up their kids from school, and the singlies enough time to get to the bookies for the last race of the day. Before venturing out they would shower.

For those who didn't go ashore, they would shower.

After sitting around for a while many of us would go and do some extra phys or the keen beans prepare lectures for the next day. Shower.

Dinner would be eaten at 1730, for us to return to the grots to watch a bit of TV. Shower.

Go ashore for a couple of beers, unless it was Tuesday, Thursday, Friday, Saturday or Sunday, then we'd go out and recreate a civilian standard stag night.

Return to camp if unable to secure a position in a strange woman's bed. Shower.

Being able to avoid dry skin, drink copious amounts of scrumpy and then being able to keep up with the fitness regime seemed to be the only criteria of making new friends in the troop. I could hold my own fitness wise and as an avid woodpecker cider drinker at school I was a more than willing participant when offered alcohol that sat warmly inside a communal plastic mug made from a 81mm mortar round casing. Steve, who I'd initially pissed off due to my squatting in his area, turned out to be my early mentor and it was he who insisted I attend the afternoon's grot party to assimilate myself into troop life.

The grot party was often the place where the lads bonded without hierarchal intervention. Here, within the confines of their own self-generated sub-culture, the lads didn't have to conform to some military convention. They could instead find themselves and be comfortable with whom they'd become. Their level of participation was only limited to their own moral values and many times guys, with their intrinsic nature to push boundaries, challenged themselves, and each other, to ensure grot parties became very messy affairs.

Grot parties allow the lads to let loose in the privacy of their own homes. A coterie of young consenting men, drinking heavily and enjoying themselves is not something new or indeed unusual and certainly not a point of moral judgement borne from the piousness of the society they serve. Here in the grots - the nuclei of all military bonding - civilian ethics are left at the threshold of normality, represented by the camp main gate, and with it, a fortress mentality burgeons. Military indoctrination can often erase their civilian past and thus blossoms paradoxical values on life - just as many civilians thrust their own ill-informed opinions on the military, reciprocal lazy stereotypes abound in the grots where forces personnel see themselves driven by loyalty to each other unlike their civilian counterparts motivated only by filthy lucre, where military people take responsibility, civilians have excuses; trust replaces whistleblowing, colour replaces blandness; comradeship replaces selfishness and heroism replaces panophobia.

I am sure it is of great disappointment to sensationalist vultures that never was I victim or witness to any young marine being bullied in these grot parties. On the contrary, they were ideal ways to introduce new lads into the bosom of the Corps. Never were we asked to do anything that others wouldn't. Some senior lads may have cheated a little to manipulate games so odds were stacked against the new guys, but participation was completely voluntary and I, along with many of my comrades, certainly felt part of the team far quicker for doing so.

My fondest memories are those bacchanalian days in the grots drinking lumpy scrumpy - the Devil's elixir that cured psychopaths of being normal. With up to ten gallons being bulk loaded into a recreation space the size of a postage stamp, 30 or so men in various stages of undress would join in and play ridiculous games, the losers suffering forfeits that would turn the hardiest of stomachs. How can one forget drinking so much scrumpy until it sprayed from my nose into the mouths of others - or sliding naked, other than a helmet and donkey slippers, head first down corridors of fire extinguisher foam, straight into nuclear-hot radiators - or drinking bodily fluids that should never be poured into my 'Big Bastard' mug?

It was here amongst the bonding of naked men, young and old, where I saw the similarity between a bootneck and the famed warriors of Sparta.

Bootneck humour is legendary. Dryer than a sun baked cracker, a Royal Marine's laconic style is in keeping with the Spartan, who were famous for their equally dry wit. It is said that during the Peloponnesian War, an Athenian messenger was sent to Sparta to warn, 'If we defeat you, we will not leave one of your children alive, we will rape your wives and take them into slavery, and none of your men shall survive.'

The Spartan messenger sarcastically replied, 'If.'

This form of humour was borne at the Commando Training Centre Royal Marines or the Spartan 'Agoge', where only those fit enough to start their training and education were allowed.

Once through the 'Agoge', young Spartans would be entered into an institutionalised relationship with an older Spartan. While a young bootneck is older than the 12 year old Spartan equivalent, many senior marines look after the young guys straight from training, attaching themselves to someone they share similar values, humour and beer brand. This pedagogic relationship is informal, where a senior marine will pass on his skills and techniques that were never taught at Commando Training Centre, just as the veteran Spartan would to his understudy.

Many a young marine seeks acceptance into the troop, as did the fledgling Hebontes Spartan, who craved entry into the mess, in ceremonies not wholly unlike grot parties. In the Spartan mess or the bootneck grot - in some would say a highly homo-erotic environment - men would hone their military skills, undertake severe physical exercise then sit around, usually naked, drinking and talking of battle, and would only leave in search of fornication, food or a fight. For Spartans, Helots would be their targets, for bootnecks, the local civvies or in times of conflict, whichever unfortunate body of men they were up against.

When duty called, Spartans and bootnecks are both revered for their accomplishments, their fearlessness in battle and their unwavering commitment to each other.

Reverence came also from certain sections of the local female population. In any town hosting a military establishment, there is a fair amount of sharing of women between the servicemen. There are only so many women and only so many men and in such a simple mathematical algorithm those attracted to servicemen may often go out with a few before settling with the right man, in an elongated 'try before you buy' scenario. Not that Taunton was a deadbeat town as similarly portrayed in 'An Officer and a Gentleman' where women tried to leave their fictional home in search of a better life. However, many Taunton girls did find the bootneck lifestyle alluring. Looking around a bootneck populated pub it wasn't hard to see why. Aesthetically pleasing, with toned bronzed bodies from yet another jaunt overseas, possibly doing heroic deeds, women found them slightly mysterious and their company interesting. Compare that to the pale pot bellied civvy who worked in Rumbelows and whose interests were solely local, venturing into another part of the country only should it be home to their favourite football team. Of course, many women preferred the sedate suburban male, so frequented the pubs where bootnecks were banned or at least kept at arms length. Yet in the smoky, loud, hot sweaty pubs of the 'Winch' 'Light Bob' 'Cellar Bar' and 'The Telegraph'; an insatiable zoo of gyrating man trappers amassed where the bootnecks congregated. These guys were, by definition, adventurous types, and many didn't seek the weight of a long-term relationship. Despite my nervousness around female company, I was there in the slips cordon, every so often some female would be nicked my way, and dependant on whether I wanted to impress her with my khaki glamour, would either end up in a conversation laced with sarcasm or genuine interest. Yet at no point did I get further than even a snog down a back alley. Despite my best efforts to get my leg over, I'd already developed a reputation for being a 'plums rating' - someone who rarely succeeds in flirtatious endeavor.

Then again, I wasn't the only onebereft of luck with females. Nick joined 4 Troop the same time as me. As 40 Commando RM was now ramping up manpower with the tour of Northern Ireland five months away, he'd been drafted in from RM Stonehouse in Plymouth. Later to be nicknamed 'Bushpig', due to his hideously hairy hog-like body, his infectious laugh and gleeful personality immediately impressed me.

Nick lived with a serious medical condition that obviously hadn't been picked up on his initial health screening prior to joining up. He lived with it daily, and although embarrassing he was quite open, unafraid to tell of the effects it had on his life. He usually suffered from it on the weekend and its symptoms were quite horrendous. It only took a few beers for the condition to appear. Once drunk, the disease would cause Nick's clothes to fall off, his naked body looking like an egg that had dropped onto a hairdresser's floor. Worse still, discarded chewing gum and a turkey neck poked from the mass of hair between his legs. It appeared to be a contagious illness - those close by would immediately feel sick.

Immensely switched on, Nick would become a reconnaissance troop stalwart. Amongst his many achievements he gained immense kudos by winning the gold medal at the yearly British military sniper competition. To be a Royal Marines sniper takes an extreme high level of competency in many soldiering disciplines and the Royal Marines sniper course is regarded as the toughest in the world, regularly attracting both domestic and international special forces candidates; so to be invited to represent the Royal Marines was a prestigious honour in itself. By winning the competition he was literally the best sniper in the whole of the British armed forces, an accolade so high that when he popped into his parents' house to tell them of the news, his mother replied, 'That's nice dear, fancy a cup of tea?'

When at the RN recruiting office, I was told by one of the matelots that blinking and breathing was deemed to be over qualified for the Royal Marines. Yet Nick himself would end up becoming a qualified electronics technician, earning a double degree in electrical engineering and applied mathematics. He wasn't the only one. Here amongst the debris of empty vodka bottles, metal trays of decaying curry and sticky women's underwear, many in 4 Troop had considerable academic qualifications. Most had 'A' levels and one chain-smoking Glaswegian had a degree in aeronautics - probably the reason why he could so accurately judge the trajectory of the piss from his penis into his own mouth.

It wasn't just academically where these guys excelled. Standards seemed to be higher in every sector of life. I thought as a successful trialist with Barnsley, I thought I could kick a football rather proficiently, but here I couldn't even get into the unit football team. Half-drunk men were consistently running sub six-minute miles in the morning and many were benching 120kg. I seemed to be in this higher plain where everyone seemed to be able to do almost everything better, quicker and harder. It became a trifle unnerving when realising that I had to up my game to reach the rarified air of elitism or be left behind as another casualty of incompetence.

'I'm a street walking cheetah with a heart full of napalm.'

~ 'Search and Destroy', Iggy Pop and the Stooges

NOW IN A REAL FIGHTING COMPANY as a real
Royal Marines Commando, my first request was to get
out of soldiering. The unit required volunteers for the
40 Commando boxing squad, so I decided to apply,
unjustly proud of my achievements when a recruit by
winning the Commando Training Centre light welterweight junior division. Bravo CSM was a little surprised.
I'd been in front of him three times now, once to join
the company and twice to leave. He gave me permission on the proviso that I would do the company proud
and win the championships. It wasn't an option.

Training was hard but enjoyable. After seeing the
unit fang farrier - the dentist who, despite me only
visiting him to fit a gumshield, decided he wanted to
conduct an impromptu extraction of all my wisdom
teeth as he was bored; we had seven weeks to turn us
from lads who could swing a comedy haymaker to
mean green fighting machines. We took on the challenge with gusto. I even learned how to skip. During
the day I trained like the professional boxers of the day:
Mike Tyson, Marvin Hagler and Roberto Duran. By
night, thanks to the lads in 4 Troop still keeping me
under their wing, I drank like other sporting profes-
sionals of the time: Alex Higgins, Jocky Wilson and the
legendary Bill Werbenuik, their logic being that drinking
heavily could only enhance my boxing skills. They had a
point. The boxing squad coach told us to keep the
alcohol consumption down to a minimum just to keep
our weight down; however, with my inability to drink
lots I found I would spend the weekend downing more

scrumpy than my scrawny body could handle, thus dehydrate through vomit, piss and diarrhoea. Monday morning weigh-ins always proved I was lighter than the previous Friday.

With the Corps Boxing Championships upon us, I felt the same nervous apprehension as I had the previous year boxing for CTC just before being refused competition entry due to being under age. This time I was old enough and I had the green beret, but still had to fight fellow bootnecks. I felt anything less than a win would seem a scar on my manliness. I was pasting myself onto a target to be sniped at should I lose in the early rounds by those who thought my only motivation for boxing was to elope from my basic duties. Ensuring my mates back in Bravo Company wouldn't take the piss out of me was motivation enough to at least win one bout.

Someone upstairs must have been looking after me. All my opponents, seemingly named Marine Rice Pudding or Corporal Paper Bag, always got in the way of my left cross, capitulating like a house of cards as they did so. Every time an opponent fell to the ring floor it truly surprised me. Whether they didn't really know the rules of boxing, or closet narcoleptics I don't know, but they chose not to get up within the count of ten. Hoofing - it meant I wouldn't have to get hit by them again. This pattern continued all the way to the semi final, where I was drawn against my doppelganger in basic training, now at 45 Commando. It was actually difficult fighting someone I had a genuine like for. He must have felt the same, as he didn't lay a finger on me. I was thankful that the couple of blows I did land put him on his arse. He got to his feet at the count of seven. He looked shaken.

'Moo gokay?' I mumbled through my gum shield.

'Mm gm moom,' he answered equally incoherently. I translated it as, 'Yeah, I'm good.'

So I put him on his arse again.

I hoped the referee would stop the fight. I didn't want to hurt him anymore. Thankfully he did. With a place in the final secured, I returned to 40 Commando rather a reluctant celebrity within Bravo Company.

Chas, my coach, instructed me to take a day's rest in preparation for the finals night. Reading the previous month's issue of the Royal Marines 'Globe and Laurel' magazine, I turned to the sports section. My opponent in the final had made a couple of entries. One article celebrated his victory in the 42 Commando cross country race; the other of him battering to a close death any opponent that had dared face him in the 42 Commando Boxing Championships. Shitty death, I was fighting Steve Ovett with a severe anger management problem. It did nothing to ease my apprehension.

If CTC boxing night was popular, the Corps Boxing Championship finals night was bigger than Ben Hur. As most commando units in the South West are relatively close to Lympstone, each unit brought along a huge contingent of support, 40 Commando seemingly the largest. We'd been pretty successful in the qualifying rounds and four other bruised men had reached the finals, including Pat who would hit national headlines years later for doing something far more impressive than punching out half of 3 Commando Brigade's middleweights. Like a scene from 'The Hulk', Pat had managed to lift up an overturned one tonne truck from a swollen river in Afghanistan to save the life of a trapped and drowning marine.

Nevertheless, other units had their fair share of bloodthirsty bootnecks who packed the gym now reminiscent of Madison Square Gardens. High-ranking dignitaries were invited as was Terry Marsh, a former Royal Marine and current world welterweight boxing champion. His ties with the Corps were still strong and called himself the 'Marine Machine' when fighting professionally.

Down in the bowels of the gym the atmosphere in the changing rooms was one of a quiet unease. While we had to feel confident, everyone would feel some form of nerves and I repeatedly told myself this to repel my own fears. I just wanted to get out there and fight, not only my opponent, but also the insecurities of inevitable battle. Thankfully, our weight category was early in the schedule therefore less time was spent feeling as if I had a prolapsed rectum.

I walked into the spotlights of the gymnasium. A huge roar greeted us. This was how it must feel to be a pro boxer. There would be few venues in the country with a larger and more voracious crowd. I stepped through the rope, catching my leg on the bottom rope to add inadvertent comedy to my entrance. I looked at my opponent in the flesh for the first time. I realised I'd forgotten an important piece of equipment - a stepladder. Standing well over six feet tall, he was a beanpole of sinewed muscle. I'd never fought anyone so tall; my only hope was that there was a lack of oxygen at his altitude. How could he be only 63.5kg?

With my arms aloft punching high into the air at a very odd angle, we battled through a first round that saw my shoulders screaming and my lungs blown. In contrast, my opponent wiped a slight bead of sweat from his unfettered brow.

At the bell to signal the end of round one Chas, my trainer and corner man, sat me down ripping the gum shield from my mouth and squeezed water into my face. 'He's fitter than you Timey.'

'Thanks for stating the bleeding obvious.' I replied, blood now apparent as I spat into the bucket.

'You're gonna have to knock him out this round,' he advised.

So I did.

Again the jabbing beanpole came at me but with his chin knocking into my left fist, he, like those before him, fell on the floor. He groggily rose to his feet, not looking in particularly good shape, so with the sportsmanship of an East German scientist, and scant regard for his well being, I again launched with a tired tirade of flurried fists. Enough, thankfully to convince the referee that he was in some degree of danger, so stopped the fight. Thank god for that, I was totally hanging out of my ring piece.

At 17 I'd become, surprisingly, the Royal Marines Light Welterweight Boxing Champion.

40 Commando won the event as the most successful unit and Terry Marsh brought a lump to everyone's throat as he gave an emotional speech about what being a Royal Marine meant to him. He was tearful as he was presented with the Commando Medal he'd won in training, but subsequently mislaid. More satisfying personally was the lads within the company were mightily proud of my achievement and as someone new, I felt humbled to feel part of their interests.

Now considered an elite within my sporting field I was hauled away from the unit to join the Corps boxing squad where my predilection for arseing about would have to be curbed somewhat.

***

40 Commando was about to embark on a month's deployment on 'Exercise Purple Warrior' which was immediately renamed 'Exercise Purple Bell end' - a

combined 3 Commando Brigade, and 5 Airborne Brigade exercise.

A month away in the cold wet and miserable weather of Western Scotland and the Isle of Arran meant they needed all available kit. Being a double yolker, I happily handed out, with little aforethought, all my equipment to supplement theirs including magazines, water bottles, extra clothing and new '*Gore-Tex*' waterproofs. I wished them well as they departed; the coldest I would suffer would probably be from an early morning swim in the English Channel. I would, however, return to a warm bed each night.

Arriving at Eastney Barracks to join the Corps boxing squad I realised these über hard sportsmen I was now amongst were all deservedly there, meaning so too was I. Yet like many times before, the limpet of trepidation glued itself to my psyche. I felt apprehensive amongst such people, inwardly seeing myself still a scrawny child.

The Corps boxing squad trained hard in the historical grandeur of Eastney Barracks. Home of the Royal Marines Museum, its operational purpose as HQ of the Major General Training, Reserve and Special Forces was soon to be obsolete, so with the MOD preparing it for private sale for housing development, it had little in the way of facilities for servicemen. The galley food was fantastic as it had so few to cater for, however, it became a culinary Tantalus, as we could hardly eat on account of having to control our weight, apart from the super heavyweight guys who would gladly wolf down a whole side of cow and then delicately boast of their fullness through the celestial aria of interminable burping.

The decrepit war-era gym had that wonderful, memory-prompting odour of stale sweat and perpetually polished wood and became our retro arena of old fashioned pain. Our long runs along the pebble beaches of Southsea, like the commandos of old made me feel I was featuring as an extra in the film *The Cockleshell Heroes* and our energy sapping training made us even fitter - fitness enhanced by the plethora of nurses at the nearby halls of residence known as 'Tampax Towers' who often coerced the other lads, already knackered from a full day of training, back to their rooms for a session of bedroom gymnastics. As per usual, my invites seemed to have been lost in the post.

We all, to a man, went into the Royal Navy Championships full of confidence. As usual, the Royal Marines were the favourites, and rightly so. Our squad contained, amongst others, two guys who later that year would reach the national ABA finals, one marine going on to win a medal at the next Commonwealth Games, and a couple future professional boxers. And me.

We'd fight sailors from the Naval commands of Portsmouth, Plymouth, Fleet Air Arm, and Royal Navy Scotland & Northern Ireland. Our trainer always maintained we'd won round one already, as the matelots often had 'green vest syndrome' - a genetic fear of bootnecks in the boxing ring. If this was true it was also true that the general rule was if a bootneck fought a matelot, then the bootneck would have to knock out the sailor. If the bout went the full three rounds and be decided on points, then the sailor would win. I never thought this could be true, but did sense when any fight reached three completed rounds, the victor was always a matelot.

I continued through the competition in confident mood, naval targets fell when hit, and my semi final bout continued in the same vein. It was against a Fleet Air Arm sailor who managed to get up twice from the canvas after carelessly running into my gloves. A third occasion of being knocked down would constitute an instant stoppage, so I was intent on putting him down one last time. I cornered him with victory bulging through my demonic glare. Lowering my head at the same time as he raised his, we clashed. I clubbed him again with a clean shot that surprisingly didn't put him on his arse. The referee stopped the fight, he'd clearly seen enough. My elation was short lived. I followed the ref's eyes to look at my green vest. As my red/green colour perception was first class I could see it was now all a bit red. Down my front ran a stream of blood. My eyebrow had been sliced open in the clash of heads.

My first ever loss. Against a matelot. I was sickened.

I trudged away to the sickbay to get my eye stitched up by a sadistic medic. Apparently as a boot-neck, I was tough as they come and didn't need any anaesthetic whilst receiving five stitches. It was nice of him to say so, but I would have still preferred some.

Like Charlie Brown, I walked under a black cloud for the rest of the day. Sometimes life just seems so shit that you think things can't get worse. Well toughen up Buster it can. A signal telegram arrived stating I had to return to 40 Commando to be deployed on Exercise Purple Warrior within 48 hours. That would be an interesting proposition considering I had no kit.

***

I reached a barn in the windswept wilds of Scotland with the glamour of an unkempt motorway service station the only highlight to a tedious nine-hour coach journey. The heavy rain that had accompanied us the couple of miles from the bus to the barn had now decided that, as we were going indoors, it didn't need to annoy us, so stopped.

Now dark, the barn shone a dim, atmospheric light from the gas tilly lamps hanging from its beams. The CSM was surprised to see me.

'Time? What are you doing here?' he asked, enclosing the second slice of bread onto a rather delicious smelling egg banjo.

'You sent for me, Sir,' I responded with the persistent blink I'd now developed from the loose stitches flicking my eyeball.

'Have I? Well unless I have had a lobotomy, I can't remember sending for you. And I don't think any of the company's officers have either. Bit of a bummer eh?'

'I suppose, Sir.'

'Did you win the Navy champs?'

'No, Sir,' I replied a little miffed. 'I got a cut in the semi finals.' I pointed to my eye just in case it wasn't clear.

'By a bootneck?'

'No Sir, by a matelot.'

'Fuck me, no wonder they've sent you up here. It's a punishment for losing to a bleeding matelot. Never mind, the company is in a harbour position round the back. Watch it out there, it's fucking freezing and it's been pissing it down all day; you'll be up to your tits in mud. Other than that, it's lovely. Now fuck off, my egg banjo is getting cold.'

I was already wet through, with little kit, and a bruised iris forever tickled by loose stiching. Dark, with no clue of anyone's whereabouts, via many irritable comments, I eventually found my section. Matt, the section 2ic allowed me to share his bivvy to prevent unnecessary noise. I unraveled my sleeping bag just as the heavens again opened in a way unique to Scotland.

Bravo Company TQ was not a happy Hector. My sleeping bag, that when I first entered felt like a hug from a large breasted supermodel, had turned out to be an incontinent old hag - it was sodden. Trying to squeeze under Matt's bivvy had been a bit of a folly and, as a result, I'd lain out in the rain for the night. I'd been told to ask the TQ if he could dry it out so I wouldn't become a liability. My first night in the field as a sprog with my new company and I was already fucking up. The TQ did wonder why I didn't have a bivvy bag. I replied that I hadn't yet been issued one as I'd been on the boxing squad.

Gore-Tex bivvy bags were the new big thing. Recently introduced, they were expensive and not to be lost unless you wanted a large bill from the avaricious QM. So expensive in fact, that the powers that be hadn't brought any extras, so it looked as though I was going to be a little 'damp' for a lot longer. Fortunately, this phase of the exercise was coming to an end so after only a couple of days of a saturated sleeping bag we returned to HMS Illustrious to sail to the Isle of Arran.

It was the first time I'd ever been on a Royal Navy ship and the aircraft carrier I now rolled upon seemed mightily impressive. Below the flight deck, 40 Commando was housed on a sea of green camp beds, the constant low hum of the engines taken over by dialogue of the film 'Highlander' playing on a continuous VHS loop, the clunking of weapons being taken apart for cleaning, and lots of Norwegian shirt-wearing bootnecks mooching around undertaking personal administration while listening to the universal soundtrack of 'The Blues Brothers' - adopted as the unofficial Corps film. What was further noticeable as a clueless sprog, was the almost permanent sound of laughter. Never from the same person, but each small group would be involved in some form of jocularity, causing a continuous ripple of mirth around the cavernous hangar. Forever cheerful, a bootneck will find humour whatever the situation. I was certainly happier. One of the lads had suffered an injury on the first phase so couldn't take any further part. Kindly, he offered me his bivvy bag and the rest of the kit needed, of which mine was currently spread around the troop.

My first real 'O' group led to my first real battle prep, that led to me lining up in my first real helicopter chalk. We knelt on the flight deck floor that had been lowered deep into the hangar. Slowly rising like something out of *Thunderbirds*, the thunderous crescendo of helicopter engines and rotor blades cutting through the dawn air hit me as I set eyes upon the transitional sky, its clarity emphasised by the biting cold. As the platform came to rest in line with the flight deck I now felt I was in a recruiting brochure. A line of whirring Sea King helicopters, their engines creating a heat haze across the flight deck, sat waiting to transport the rows and rows of kneeling commandos from this mother ship across to where I now looked - the outline of Arran, a spectacular backdrop that cast its beauty across a shimmering sea.

With a permanent smile that only a child can exude, I sat in the helo watching eagerly as we flew in arrowhead formation to the landing site. Disembarking, we crouched in a huddle over a pile of heavy bergans until the hurricane force winds and the whirring of the rotor blades faded into the distance. I looked ahead. I was transported to the images of the Falklands war, a snake of commandos heavily laden, yomping over rolling, barren ground where tussocks of grass were the only things brave enough to grow.

We yomped for endless hours with our thumbs shoved firmly up arses, the only sound the thud of someone losing their footing and the following expletive and the laughs of others. While just deemed a break from the monotony of the yomp by some, to me I was immensely excited when eventually I was to be involved in my first commando attack. The Troop Sergeant gave us quick battle orders to 'neutralise the enemy' that had been observed ahead in the thick freezing fog. We advanced hurriedly to contact before a section corporal spotted some movement ahead. The fog gave excellent cover, but we still dived into the hollows of dead ground while orders were given to attack the position ahead. On command we launched into an attack, perfectly executing the drills guys had done so many times to an enemy that didn't return fire. The attack came to an abrupt halt. We hadn't attacked the enemy. We had been attacking a flock of sheep. Cowardly sheep at that, they didn't even stand and fight, but retreated into the foggy depths. I do recall one of the lads in the troop twisting an ankle so, in effect, in my first ever commando attack we came off worse against a flock of gutless sheep.

Carrying on to our final objective, we immediately swept into clearance and standing patrols. Even with the slower pace of a large-scale exercise such as this there was little time for resting, and as the rain decided to persist, I felt strong, committed, and eager to carry out whatever was to be done. I had become oblivious to the harshness of the environment. I could laugh while piss wet through; I wasn't treated like a sprog here in the field, I was treated like a member of the family.

It was my 18th birthday. I sat with my coffee and whisky, kindly donated by Matt in honour of my graduation to adulthood, listening to the rain drops filtering through the pine trees above, reading a Dear John letter - a note to say I'd been dumped - from

Jayne, a girl I'd experienced a frustrating fling with whilst on summer leave. I realised that she didn't really matter. While I had a great affection for her ours was undoubtedly a meretricious relationship - my one visit in the previous eight weeks the only proof needed. I was now amongst people with whom I'd have a lifetime relationship. Whereas in training I'd struggled just to get through the hard times, I now welcomed them, I knew I had brothers alongside me who'd been through the same shit, people who understood the meaning of unity, brotherhood and teamwork. These were colleagues who would look after me with dark humour then take the piss when I read out my Dear John, friends who would then make sure I was OK. They did as I would do unto them. It took that moment, with perfume scented pink notepaper in hand looking over the bleak beauty of Arran, to realise I'd finally developed into a commando.

***

40 Commando RM returned from Scotland and I was greeted with an 18th birthday present from my mother. I now have a teenage son, so I know the difficulties of finding something appropriate. We look at what his interests are, try to recognise hints, or follow his lingering gaze when browsing the shops. I guess my mother and stepdad didn't use this type of plan. I was 18 years old, in the Royal Marines living in barracks, single with only a brief fling as my sole relationship, interested in music, boxing, alcohol, and had designs on travelling whenever I could. Using these parameters my parents came up with the perfect present:

Clothing? No. Driving lessons? A celebratory bottle of champagne? A gift voucher, offered when effort is lacking? No. No No.

All of the above, while not perfect, would've been gratefully received.

Instead, here I was, a single 18 year old commando opening my present to find…a knife… an electric fucking carving knife.

Gordon Bennett, what was I to do with an electric carving knife? Would I attach it to the end of my weapon and use it as a new fangle, but totally useless bayonet? Did my mother think I could plug it into a tree and use it as some form of commando dagger, instead of cleanly severing a carotid artery I could slowly carve off an enemy head? Or could I use it as something to fill the dustbin? Apparently, she bought it for me to put in my 'bottom drawer' in anticipation of a wedding. A wedding? I didn't have a girlfriend. I didn't even have a drawer handle, never mind a bottom drawer to which I could attach it.

***

Like Christmas, I'd waited eagerly for its arrival - preparation for the forthcoming Northern Ireland tour was now upon us.

It was rather ironic that lads would parade in the morning to train mercilessly for a peacekeeping role then go out at night and be involved in anything other than being quiet. Incidence of partying seemed to increase in direct proportion to the intensity of pre-deployment training.

Also of contrast was my mood. I'd often found myself unhappy for extended periods. It wasn't a worry, everyone becomes unhappy at times, but more of a confusion. Why would I feel this way when life was so good? It was nothing new. As a child I would often turn from a smiling cherub into a scowling imp. This could be passed off as childish sulkiness but now as a man I thought I'd have grown out of it. Sleep was often my cure and lethargy could be passed off as feeling a little ill. 'The body', I argued 'is nature's own doctor.'

Yet when the devil of darkness crept back into one of my psyche's filing cabinets, I would wake as if someone had dropped an effervescent tablet in my mind to rewire then happily go ashore with the lads and offer myself as a sacrificial source of tomfoolery.

Knowing we were soon to be on operations, the 'brotherhood' vaccine we had been given along with Hep B and Typhoid was further coursing through our veins. As a troop we would go out in a large group, only breaking up late at night should someone get lucky with a 'Doris' or decide they preferred a kebab to embarrassment on the dance floor. We trained together, we ate together, we drank together, and we would fight together. We were close. So close, that the only possible thing that could bring us further together was a good old game of spoof.

Spoof is the official game of the Royal Marines. Should you ever question the veracity of someone's claim they are or were a Royal Marine; offer them a game of spoof. If they decline or don't know what spoof is, they are clearly a 'Walter Mitty'. It is programmed into a Royal Marine, just like a dog to chase a ball, to play spoof.

It is a game often played in a drinking establishment. The rules are quite simple. A group stands in a circle and each player hides three coins behind his back. When a round is played, each player brings either one, two, three or no coins into the centre hidden within his fist. The challenge is for each player to guess how many coins are hidden in the fists within the circle. The one who guesses correctly is out of the game. The first player out has the privilege of setting the forfeit for the last man remaining in the circle who is considered the loser. The game has military validity. It is a game of analysis, and uncertainty. There is a military saying that 'no plan survives first contact', and spoof ensures that any plans made can be scuppered by another player's decision-making, requiring a player to constantly think on his feet. As more players leave the game the demand builds on those left creating an example of how guys react under increasing pressure. The forfeit for the loser can usually be a round of drinks for the players or when the ante is upped, a ridiculous tattoo of such things as the NAAFI opening times or a hoover.

Variants of spoof could be added, non-emotional spoof was an often tried and tested version when the initial excitement of calling the correct number of coins had to be tempered with only the words, 'Thank you gentlemen spoofers,' as an acknowledgment of victory. Anything more apparent, such as the corner of the mouth being raised or a deep breath, were clear indications of emotions and the player would be launched begrudgingly back into the game. Another variant was over-emotional spoof where the player out of the game would have to undergo some sort of over exaggerated banshee-like seizure doing cartwheels, star jumps, or running to the toilets grabbing some toilet blocks in the urinal and chomping them in front of the rest stating they were lemon sherbets. Either that, or just sit and piss his trousers creating a map of Corsica in his jeans. Over emotional spoof was best played in a quiet pub on a Wednesday afternoon.

So, on the occasion where Matt lost and his forfeit was to be blindfolded by a pub's beer towels then, crawling along a line of 17 unzipped men, sniff each penis as he went along to check who had the smelliest bell-end was of no real significance to us. Apart, of course, from the honking creature deemed to have the smelliest helmet. If it had been done in the privacy of the grots on camp it wouldn't have really been an issue, but as it was in a pub, where two lads decided to pour themselves and the rest of us free pints, it became public knowledge. With members of the company already infamous for waking up people by bouncing their erect cocks off a sleeping man's forehead, Bravo Company became 'Beef Bravo' in reference to being gay. It was a totally unfounded accusation. We weren't gay; we were just friendly.

'Maturity begins to grow when you can sense your concern for others outweighing your concern for yourself.'

~ John MacNaughton, Author

NORTHERN IRELAND TRAINING to me was a totally new experience, a different form of warfare not really covered in basic training. We now trained in four man teams rather than eight men sections. As a quartet, we would perfect our movements and positions like Olympic synchronised swimmers, each of us instinctively knowing the actions of other team members, irrespective of whether we operated in an urban or rural environment. As our tour was in the 'bandit country' of South Armagh we'd have to be proficient in both. We were educated on the basic history of the troubles, and who were the main 'actors' ('enemy' is such an aggressive name for belligerents in peacekeeping; God forbid we offend anyone in a war). This ended up being another glossary of abbreviations: Provisional Irish Republican Army (PIRA), Ulster Volunteer Force (UVF), INLA, UFF, RHC, DMSU, RUC. We learnt specialist surveillance and intelligence gathering techniques incorporated into a build-up programme prior to the main training course at NITAT (Northern Ireland Training and Tactics) centre in Kent, now called OPTAG due to the prevalence of other conflicts.

The centre is something out of a 007 movie. The whole place is set up like a film set and all potential threats and our responses to them were practiced over and over and filmed for our viewing pleasure afterwards. This militaristic bravura, it had to be said, impressed. Our team drills were like KY Jelly on a deviant's dick - smooth, slick and prepared to go anywhere. Our three metre checks around lampposts and street corners were on the ball, looking for tell tale signs of IED wires, disturbed masonry in which an IED could be hidden or dog shit. Why three metres, I don't know. It didn't give me confidence that I'd survive a head high bomb planted four metres away. We took part in live riots, men from 42 Commando playing the part of rioters throwing petrol bombs, bricks and anything else at us that was large and heavy with the capability of being fashioned into a missile. If there had been a local travelling dwarf troupe I'm sure they'd have been covered in broken glass and hurled over the top of our shields. Of course, we'd respond by forcing them back in our Spartan-like shield manoeuvres, and would open up now and again to fire off blank baton rounds. I say blank baton rounds, one young lad, Ray, in a moment of delirious foolhardiness decided to place an actual baton round down his launcher. The baton round had been thrown by someone from rent-a-mob, so he thought returning it the most courteous thing to do. Whether he knew it or not, but placing it down on top of a blank cartridge basically made it a live baton gun, so when given the opportunity to fire, he shot at a Naval WREN who was annoying everyone with her crow-like screaming at the front of the crowd. The round fizzled through the air, breaking her arm. It wasn't an act that would win him promotion. Unfortunately, it didn't shut her up. Her screams of abuse turned to screams of agony.

The Special Investigations Branch did investigate Ray and was subsequently charged but he did deploy. I don't think he was given the baton gun though, despite proof that he had a decent aim.

The training was exceptionally realistic, as the broken-armed WREN could attest, and I learned a great deal in those weeks and absorbed the information like a sponge. I must have impressed somebody, whether the Troop Sergeant, Scouse, or my Brick Commander, Les, as I was given a few important roles despite my inexperience.

Regardless of the fact I'd built up a reputation as someone who was generally shit at dating women, I was handed the role of 'chat up man' at the vehicle check points (VCPs) that we'd man while on patrol. I'd be the one to stop the car, make first contact using a friendly but firm approach to the occupants and search the vehicle if necessary. It was a role I took to like a duck to water and soon found myself able to naturally talk to drivers without some ulterior pretence. I was also blessed with a near photographic memory, so when shown the rogue's gallery of paramilitary identities, I could recite most almost immediately. At last, I had something to offer - a great tool for intelligence gathering.

As the training wore on and the more familiar I became with our aims, the more I realised we were just mobile targets for the evanescent PIRA volunteers. Patrolling the fields of Kent only proved our point. Our days consisted solely of reacting to enemy attacks, our role purely defensive. The chances of us catching anyone planting a bomb or pointing a weapon at us were virtually nil. The chances for us to fire first were infinitesimally small. The odds were stacked innumerably in PIRA's favour. I vowed to ensure I lost my virginity before some Republican could pop a cap in my ass before I popped my cherry.

***

We flew into RAF Aldergrove at night. Most noticeable when coming into land, was the inescapable neon copses of telecommunication towers flashing in and around Belfast - it was clear we were in no ordinary part of the UK.

Nick had told me a tale of when his father arrived in Northern Ireland in the late 1960's arriving by air before boarding a 4 tonne lorry to drive down to Bessbrook Mill, with only a sentry at the tailgate carrying 10 rounds for protection. How ludicrous was that? That was in the 60's, things had moved on. It certainly had. We now boarded a 3 tonne lorry and the sentry had 20 rounds.

We sat silently, all aware of our vulnerability as the night convoy moved along the damp, unlit country roads of Northern Ireland until we reached our destination - the small village of Forkhill near the border with the Republic of Ireland. I was now on my first operational tour, maybe it was my youthful naivety but I didn't feel scared; my overriding emotion was one of excitement.

We quickly sorted out our personal administration before a quick intelligence briefing. Immediately after, a short yet exhilarating helo ride took us to R21, the overt Observation Post (OP) overlooking the Dromintee Bowl, an area of outstanding natural beauty and home to some of PIRAs most proficient fighters who used religion as the respectable shop front to their mafia style business. Despite many of us being non-believers, we had been cast into a Theo-political purgatory, where local Catholics would see us as demons and British politicians used us as cherubic pawns, with the return to the heavenly delights of normality our reward for professionalism and survival. We'd entered a divided land where, depending on the street I walked, someone could want to hug me or torture me due to the clothes I wore. No longer would we have routine, promoted within the cossetted safety of the mainland, as setting patterns made us easier targets. This was an unyielding place where danger lurked around many corners, where a quiet street could be the precursor to a bomb, where a watchful young lad on the street could be a messenger to someone intent on causing us harm, or a stranger's random conversation could be the stalling to ensure a sniper was properly positioned. It is unfathomable, that when entering such a dystopian world, young men and women return to the relative utopia of society without any mental scarring. It is totally understandable when they do.

Our arrival had not gone unnoticed. IRA's proficient intelligence network meant they were fully aware of our arrival. In their appreciation, instead of putting out the bunting, they asked two volunteers to welcome us with a far more thoughtful present - a van full of explosives. Unfortunately for the two volunteers, while preparing the Improvised Explosive Device (IED) it detonated prematurely killing them both. A bit careless really. We saw photos of the two men in their prime, and how they looked after the explosion. In the 'before' shot one looked like an accountant, the 'after' shot he looked more like dog meat. The other volunteer had looked quite short in his photograph but now he had grown to around a 1000 metres, his body parts spread across the countryside. For me, as an 18 year old target, not one fuck was given for their demise.

The news, although welcome, was sobering. There was an enemy out there, no matter how unseen, and we were the obvious targets. PIRA wanted to hit someone with a high profile and a Royal Marines Commando would do just fine.

A detailed handover by the Scots Guards complete, we were now fully operational having control on the ground. The Scots Guards had finished a successful tour and seemed a professional bunch, and our advance party had been made to feel most welcome even if they had to drink lots of Irn Bru and wear helmets. Unlike the army units, our legendary Commanding Officer was adamant we would wear our green berets on patrol. Helmets could be seen to be aggressive, only worn in war. It also took away the individuality of the unit concerned. We could have been anyone. The wearing of berets could be seen to show that we were:

a) Understanding

b) In control

c) Commandos

And we would look a lot cooler if shot.

Due to my memory skills and unerring ability to identify anyone of interest from an odd angle, I'd been chosen to sit with Scouse the Troop Sergeant, and take up the 0800-1600 watch at the OP. These 'persons of interest' were known as 'players' - not defined as a lothario so common today, more in the 'if we capture you, you will be tortured and dismembered' kind of way.

Having no mains sewerage on the OPs meant we had chemical toilets that had to be emptied and burnt off daily. Being the sprogs, Bryn and I would often get shit burning duties. I couldn't have been happier. Looking over the picturesque countryside below, human faeces wafting across our nostrils like 2 kids in a *Bisto* advert, we wondered how such a beautiful part of the world could be such a hotbed of violence and hatred. As we stood throwing diesel onto our shit pile, a huge explosion just north of our position broke the serenity of the landscape. We looked at each and feared the worst. We knew our guys were on the ground and only two days after PIRA had scored an own goal we wondered whether this time they had finally secured a result.

The news finally emerged. The explosion had actually come from the Royal Artillery guys who were manning an OP on another prominent feature. Some clown, obviously missing the sound of loud bangs, had negligently discharged a PAD mine sending thousands of ball bearings into the local area, luckily nothing was hurt but the idiot's bank balance.

***

Our lives, and sleeping patterns were dictated by a three-week rotation. One week in the OPs, a week on rural patrols followed by a week combining sangar duties and town patrols. The OP week was the one I relished. I knew I was an asset up there sat in a box, and life was simple. We were fed well as we had our own chef, we knew what we had to do, and we did it well. As I had the business hours shift I had plenty of time in the evening to exercise, read, and burn shit. I was finally living the dream.

Rural patrols tended to bring the more natural soldiering out in us. We would patrol with immense weight, carrying all manner of electrical 'bomb stoppers' to prevent radio controlled IEDs being detonated. We could be on the ground for anything up to a week and the hyper-sensitising of vulnerability overshadowed any feeling of isolation should we get hit.

The rotation that most of us wanted to end swiftly had to be the week on camp manning the sangars. A sangar, for the uninitiated, is a fortified sentry post usually on the perimeter of a camp built to raise early alarm. It could be positioned on the ground or preferably in the air, somewhat like a lookout tower, to create a better vista.

Should you want to get the sangar experience the only way to replicate life on such a duty is to get a large saucepan or colander and put it on your head. Go collect a mop/sweeping brush/swimming pool net if you are rich enough to afford a pool (in which case you probably won't ever do a sangar duty). Wait until dark, then traipse down to your garden shed, then for two hours at a time stand up looking through the window at the blackness of your surroundings. If your neighbours don't report you to the police, try and do this for 24 hours. In your two hours off try to sleep or eat egg sandwiches and coffee, whinge about being on sentry duty, and complain that the pornography in the sangar is unreadable as all the pictures have now stuck together. Then go to work. You will feel more complete as a person I am sure.

To those with an active mind, sangar duties were a stoppage in time. Clock watching became an Olympic sport but the large hand would seem to progress at a far slower rate than usual. Was it the blurred luminosity of the hands that made it seem so or was it just looking through blurry eyes? Imaginations rioted as time allowed us to permutate the league tables in the upcoming football fixtures, evaluate the weaknesses of the internal combustion engine or plan a calculated assassination of any IRA local. To those happy to stand gormlessly looking out, it was also a great opportunity to get some peace while partaking in frenetic masturbation, poring over the porn magazines hidden behind the bullet proof plated walls. In such a popular onanistic emporium, some of these magazines could only be handled with gloves, a pictorial health hazard, so well used they had turned to hardback editions. While the sangars were non-smoking, the few nicotine addicts would have a crafty one ensuring the glowing embers were invisible to those outside.

Permanently manned by sweaty bodies, a 24 hour rotation of smoking and wanking left the sangars not a place from where you would write a postcard saying 'Wish you were here' unless your friend happened to be an agoraphobic chain smoking sexual deviant.

Fortunately, these days of abject boredom were interspersed with short town patrols around the village, usually done to prevent any attack on the base. Our patrols seemed to catch the eye of the locals. I know we were a good-looking bunch of bastards but the stares we received were more than we had bargained for. An accompanying RUC officer explained that our new weapons were a source of curiosity to the locals. We were the first unit ever to go to Northern Ireland with the new SA80 rifle. Previous units had used the old SLR, the weapon I had used throughout training. We were here with Britain's newest and finest piece of weaponry. So fine that the sights kept fogging up, parts kept dropping off, screws constantly loosened and its 5.56mm calibre meant to kill anything larger than a rabbit you'd have to shoot it repeatedly. The sling system was a conundrum of webbing and buckles that just asked to get caught in anything remotely bush-like and totally inappropriate for the many hedgerows of South Armagh.

The TQ must have been watching *'Chitty Chitty Bang Bang'* in his down time, as he suddenly became the reincarnation of Caractacus Potts, trying to invent anything that would make the SA80 a little bit more efficient. Unfortunately he couldn't turn it back into an SLR or an M16, so instead buried himself in his store next to the helicopter landing pad to try and adapt the weapon into something useful.

The magazine release catch was a major problem. In the Province losing a magazine created immense problems. Should the magazine fall and not be noticed then a search operation may have to be launched to recover it safely. PIRA had a habit of booby trapping lost or left behind equipment so to recover anything after a noticeable time period required a range of military assets at considerable burden to the already overworked units involved. Of course, the Security Forces could exploit this ploy, as a discarded magazine dropped accidentally or otherwise could be surveilled and intelligence gathered.

After a few days of tinkering, the TQ exited with an air of triumph having cut, spliced and reconstituted the sling so instead of wrapping us up like an angry octopus could be used efficiently and swiftly in any form of carriage.

To solve the problem of the magazine inadvertently dropping off, he engineered holes through the bottom of the magazine through which we threaded string (the hoard I kept after training did have its uses), which we then attached to the rear sling swivel on the weapon's butt plate, meaning if it did fall it would just swing from the weapon like a mitten from a child's duffel coat. Of course, if we were in a firefight we would have to change magazines with one perpetually dangling and swinging around and one in the magazine housing that could drop out at any moment. He eventually managed to get the armourers to make up small metal guards to protect the makeshift magazine release catch but still retained the string as trust in our weapon had yet to be earned.

The upside of this new weapon of moss destruction (as moss was about the only thing the weapon's original variant could actually kill) was we could fool the locals with any old bullshit, including that it worked. While on town patrols, I taped my thermos flask underneath the front hand guard, telling the inquisitive kids it was a grenade launcher. Not only was this a little joke, but we knew that these innocent kiddies would go blabbing back to their mammy and daddy about the impressive arsenal we now carried daily.

Bullshit does work in a dirty war such as this. The PIRA members were on their home turf and so we had to use any trick in the book to gain an advantage.

Nick was given the team leader position for the day, Les, my team commander, was injured so we shared the leadership duties, bringing along one of the HQ guys to make up the quartet. In my normal role as chat up man I had the fortunate task to stop four of the big time 'players' at an impromptu 'snap' VCP. We didn't often see those high in the PIRA rank strata, they were usually far too clever to get caught, letting their donkeys get shit from us bastard 'squaddies'. The driver did as he was asked; he knew how to play the game. The passenger; however, was less diplomatic.

'How oild are yuy sonny?' he asked with a Northern Irish accent as thick as cold treacle.

In fairness, I must have looked about nine.

'Does your mammy knoy your oyt? The other occupants laughed at his banter.

Here I was getting grief from these PIRA stalwarts, how would I respond? Nick, who stood with me witnessing the banter, thought the best form of defence was to politely request their ID. We knew who they were, they knew we knew who they were and we knew they knew we knew who they were, so to avoid further confusion Nick pointed each driving licence towards R21 OP some five miles away. Holding it up for about five seconds he pretended to have a conversation into his radio.

The driver looked puzzled. 'What yuy doing?'

'Just getting your IDs on our new system, making sure everything's OK,' said Nick smiling politely.

Still he wore a mask of confusion.

'Sorry, just making sure the lads on the telescope can see the ID clearly and we'll record it and put it on our records.'

The driver looked all the way to R21, too small to see from this distance but overbearing to the people at its foot.

'Yuy can see it from there?' He now looked worried.

'You don't think NASA makes telescopes just for space do you?'

I'd searched this particular car early in the tour but here the driver didn't recognise me from Adam. I recalled he had suffered an oil leak in his boot so now as I was inspecting the boot again months later I asked how he'd got his oil stains out. He looked at me again with alarm.

'How yuy know aboyt that?'

'Aah well you'd be surprised what we know about you.'

Even just placing two little seeds of doubt over something as trivial as an oil leak or as ridiculous as the ruse of a Hubble telescope sitting in a tiny hut on a hill in South Armagh was a small victory in a huge conflict, being unsure of our capabilities meant it could cast doubt on theirs, even more satisfying when placing an 'I've met the marines' sticker on their boot as they drove off. This was far less impressive than the Support Company lads gluing a green beret to the head of the

IRA memorial in Crossmaglen. Interestingly enough, this statue was sculpted by a Nazi ally, collaborator and sympathizer in WWII; a mere irrelevance to many Republicans as long as he hated the British.

***

I went to Northern Ireland to get my first taste of action. But half way through the tour, all I had was my first taste of piss.

Embarking on R'N'R the following day, Bessbrook Mill became our temporary home. In a more relaxed environment we enjoyed drinking games that involved chewed peanuts and warm urine, neither one recommendable to put in your mouth. The puritanical NAAFI staff closed the bar in disgust. In a tit for tat show of revolting revenge Nick, who as usual was naked, shit in a polystyrene cup and together with a note saying 'cheers for the beers' cooked it in the communal microwave. The resultant explosion certainly confirmed we were in a dirty war.

R'N'R is aptly named. We could now rest and re-cuperate for five whole days. Being constantly alert becomes the norm when in an environment such as South Armagh. Like driving fast in horrendous conditions, the senses cultivate hair trigger responses that cause gradual fatigue, stubbornly fended off by the desire to protect oppos and get the job done. Only now, while relaxed in the comfort of the safe processed banality of the mainland, did I realise how tightly wound I'd been. No more did I think that my next step could be my last with my unused manhood intact. Here, I could stand by a lamppost and not worry that my legs would be blown across the road by an explosive device hidden inside masonry or receive a drastic centre parting from a well aimed sniper shot.

Since we'd been in each other's pockets for the past three months it only seemed fair we'd spend our R'N'R together, so our team decided to return to Bristol where Les lived.

All day drinking is not something I excel at. Indeed, there is a section 60 in the Army Regulations that states someone can be charged with 'bringing the military into disrepute' and my ability to consume beer at midday certainly would contravene it. So by the time I entered 'Papillons' nightclub ,I was a burbling wreck with an aura of senseless asininity. Like most nightclubs, 'Papillons' was a tumultuous bubble of sensory overload. I'd not been in the club more than ten minutes when a woman approached me. She wafted through the dry ice fog like some medieval sorceress with a fluorescent drink containing an umbrella in one hand, and a half drunk pint in the other. She was certainly prepared. I can't remember what exactly she said, I never heard her fully over the ridiculously high decibels being amplified around the club. Whatever the subject matter, she led me by the hand to the entrance past Nick who gave me a thumbs up - that knowing 'good luck' thumbs up that is seen attached to young soldiers on the war reels and posters of the 20th Century. Less familiar was the look of foreboding that was surely etched across my face.

While camouflaged by neon, fog and general lack of ambient light, she'd looked quite fashionable, not quite Milanese catwalk model standard but she wouldn't have been the only female patron to promenade her sexuality wearing a leather top with lacy sleeves. That is, if she'd actually worn one. Evidently unable to get transport to some reputable clothing retailer, she'd fashioned her own version. Her PVC waistcoat didn't have proper sleeves; rather they were nylon stockings with the toes cut out, her left sleeve Sellotaped to the shoulder of her waistcoat. While resourceful, her planning skills were somewhat limited; she'd obviously run out of Sellotape as her right sleeve was connected by blue electrical tape.

In anyone's right mind, now would have been an opportune time to disclose some feigned illness. But I wasn't in my right mind. A mix of libido and alcohol was currently coursing through my phallic veins and sending irresponsible messages to my befuddled brain.

Before my brain had registered any form of objection I was outside her bedsit in the suburb of Kingswood. I remember she was 24 years old and worked in a shoe shop, clearly not a clothes shop. Her prosaic chattering paled into the subconscious, as all I could think about was my imminent passage into my new life as a non-virgin. I was nervous. The sort of nerves that you have just before going on stage; nerves so intense you feel nauseous; or was it the beer? Either way, I didn't want to have the moment spoilt by me puking all over her back, or front depending on which way she threw me.

Her bedsit looked like the final day in Hitler's bunker, (messy, not full of swastikas) but Eva Braun surely would have not left a half eaten *Pot Noodle* lying on the bed. It was the last glimpse of my new environment before she kissed me with the force of an angry drunk's headbutt. Her tongue nearly choked me as I tried to respond like the novice I was, and tasted the nicotine in her mouth. She threw me onto the bed and stripped naked, not quite in the way that a sensual stripper does, more like a woman with her clothes on fire.

Three things scare an 18 year old male.

1. Being buried alive.

2. Out of date crisps.

3. Over-sexually aggressive women.

At this point, it would've been preferable to lie in an open grave eating a packet of two year old *Monster Munch*. I now knew how a male praying mantis feels just before a bit of rumpy pumpy. I'd been patrolling the badlands of Northern Ireland for the past three months and I'd never been so scared as now.

I also disrobed, leaving my socks on (as anyone knows, they are the first line in protection) and felt a woman's mouth on me for the first time. With my eyes closed I had to admit it felt pretty good. Better than an orange anyway.

Being the generous lover that I was yet to be, I decided to reciprocate the act and since the nearest I'd yet come to tasting a woman was licking the finger that I'd just withdrawn from the edge of a stripper's labia at my King's squad piss up whilst in the final week of training, I was keen to sample new flavours. The view from my position made her vagina look like Tina Turner eating a selection of cured meats and certainly wasn't as attractive as I'd hoped. Nevertheless, it wasn't in my character to refuse anything, even odd looking labia, so took the plunge. It was an odd taste, rather like battery acid, not unpleasant but not quite the nectar that some lads had described it.

I put my fingers in her with the sensuality of a sadistic midwife. It was only the second time I'd ventured this far, and it felt different from the stripper I'd first managed to seduce into allowing my fingers into the tunnel of love. This time it felt as though I'd put my quivering hand into a bowl of tepid mince and, rather unnervingly, made a similar sound. Thinking back to my sex education classes as school and the profile view diagram of a woman's pelvic region one could only surmise that the squidgy object that was now compressing her vaginal wall was in fact the world's biggest poo. A poo that could be used as a draught excluder (only in an emergency mind, I don't think even the most depraved of German porn directors would choose it over a long foam sausage), a poo that if someone back at Norton Manor Camp would have produced it would have been photographed and framed. In war it could have been used as a sandbag to hide behind for cover. I think you get the idea.

She was either enjoying my time down there amongst the heather of her wiry pubes or she had some sort of seizure as her legs crushed my head as she writhed about in agony or ecstasy, bending the top of my ear over until it became quite tender. Before I became bored, she thankfully muttered those two words that I had so longed to hear, just maybe from a different person.

'Fuck me.' It was not with the sensual intonation that I had hoped, more of a bank robber's demand.

I was about to partake in the ultimate act of intimacy between two human beings, an irreversible bond between two people. I'd resigned myself that my first 'bag off' wouldn't be under a swaying palm tree with the warm azure seas of the South Pacific lapping my toes, yet I didn't expect my initiation into sexual congress to be in an untidy room, only one level up from a squat, with an interloper interrupting our intimacy - the sound of persistent phlegm regurgitation from a man dying on the other side of paper thin walls.

Throughout puberty my fantasy woman had changed. The girl that now lay down under me was none of those I'd dreamily flirted with, and I suddenly questioned whether I should continue. I don't think I had much choice in the end. I doubted whether enjoyment would follow. It was like stuffing a marshmallow into a welly top .

I looked into her eyes, and wondered which one to look at as they didn't point in the same direction, in fact one only pointed forward, even when her other one looked to the side, possibly at the clock on her bedside table. As I'd always been slightly boz-eyed I wondered who was the more perturbed. I closed my eyes again and dreamt of my fantasy women I'd disgraced myself over on numerous occasions through puberty: Kim Wilde, Nadia Comaneci (still do), Catherine Bach and Jayne Seymour. At intermittent intervals I opened my eyes, but none of those women lay there, just a bored looking Bristolian with wonky eyes.

'Here, let me get on top.' She'd obviously had enough of inhaling my beer fumes and amateurish frottage creating nothing but friction burns on her legs.

I'd mistakenly thought that love making was a calm, loving affair - it'd said so in the '*Joy of Sex*' book. I can't remember anywhere it suggesting that the female when on top should pretend she was riding a stallion at the Calgary Stampede. The tempo had certainly quickened. She stopped abruptly.

'Are you OK?' she questioned, sounding identical to the hot female doctor in the act of scraping urethra cells from inside my nob, and of the two acts I was unsure which was the more pleasurable.

'Err yeah hoofing,' I lied.

Forgetting her concern, the event only took a couple more minutes. I told her I was about to 'shoot my muck,' not quite Mills and Boon but she got the message all the same. She jumped off me and took me in her mouth once more, until I emptied myself with the face of Nadia Comaneci projected onto the inside of my eyelids.

Whether she could taste garlic I don't know but she trotted off to the bathroom and swilled her mouth before closing the door. I didn't know whether now would be a good time to ask her name.

I thought it necessary to follow corps tradition and wipe my nob on the curtains but decided that blinds would be a trifle more painful, so lay there in the afterglow of my first sexual encounter listening to the splash of toilet water as she took her well needed shit. She didn't even have the decency to put some toilet paper down first.

R'N'R over, returning to South Armagh a heroic non-virgin wasn't personally seen as a drag. We were half way through the tour and, strange as it seemed, I felt happier amongst the wet, cold and conflict of Northern Ireland.

The world is a dangerous place, not because of those who do evil,
but because of those who look on and do nothing.'

~ Albert Einstein, physicist, hair model

ON OUR RETURN, the Forkhill SF base had a new member. Des the dog, a smooth haired tan Jack Russell, had become Bravo Company's resident canine hero.

Military units have traditionally taken local animals to their hearts and Des had followed a random patrol back to Forkhill barracks. After being searched he was given a full kennel name of 'Bishop Desmond Tutu of Shittington' and colloquially known as 'Des'. Security cleared, he was eventually allowed to stay with the company, and a home was made for him in the NAAFI.

His bravery was never in doubt; he often took point position at the front of the patrol, fearless as the first patrol member out of the SF base and did not shirk when approached by other dogs, gladly sniffing their arses, we reasoned, to check for explosives. After his first few patrols we felt he needed a little more protection so created a set of webbing for him, complete with pouches for emergency dog rations should he get split from the patrol. The lads treated him like a dog king and his belly was forever full of corned beef and cheese. Des was a stalwart of the company and it was one of the saddest days of the tour when a local rocked up at the RUC enquiring whether his dog was there. Des was summoned and seemed to know the man judging by his wagging tail. He must have been surprised when grabbed by the scruff of the neck then thrown unceremoniously into the boot of his owner's car. We never saw Des again and his loss thrust a blow to the morale of Bravo Company.

Now without a pet, we could even venture abroad. Les had been called away on some intelligence meeting leaving our team to be commanded by one of the signal specialists keen to get out of the Ops room. Banksy was a great guy and was happy to learn the ground from those of us who'd trodden the area many times already. Tasked with hand railing the border, this was true bandit country where a highly professional PIRA sniper was now taking out targets in the North from his position in Eire.

After patrolling through the endless thick woodland, we went to ground as Banksy carefully studied his map.

'Right lads, we're near the border. When we hit a stream let me know, as that's the demarcation line.'

Nick looked at me. I looked at Bryn. Bryn looked at Nick. Who was going to tell him?

'Banksy, we passed over a stream about 5 minutes ago,' said Nick.

'Did we?'

'Yup, Bryn's got a wet foot to prove it.'

Bryn lifted his wet left boot in confirmation.

'Fuck, I never noticed.'

'So much for us being amphibious specialists,' I said.

We left the tree line onto a metal track and found road signs definitely written in a different style and language.

'Welcome to Narnia,' said Nick. 'At least we're safe here.'

'How do you make that out?' asked Banksy, now in a bit of a flap.

'Well that sniper has been firing into the north. He'll be facing the wrong way to have a shot at us.'

Invasion over, we returned to our rightful country without any immigration checks or duty free *Toblerones*.

***

Patrolling the badlands of the South Armagh country-side our paths consistently crossed a fearsome foe. Not PIRA, no. They were too busy hiding, plotting and thinking of evil things over a cup of Yorkshire tea and a packet of digestives.

We came up against something far more formida-ble - the blackthorn hedge. Historically planted to split up familial land, they were evidently created by the gods to prevent any mortal accessing the heavens without first overcoming the mesh of thick teeth-like thorns sat upon sturdy branches that could rebound with enough force to remove a child's face. Anyone able to cut, slash and worm their way past these hedges would be worthy indeed of reaching heaven and therefore attain god-like status.

We could have used the gates like the farmers, but PIRA volunteers, after dropping off their kids at school, booby-trapped gates for fun, so piercing the dense hedgerows was the only safe way to advance. PIRA wouldn't booby trap a hedge, watching the lads fight against these barbed hedges was too good a sport to watch.

Picking a spot that looked accessible I'd push, only to be pushed back by the security guard branches at the front. So I'd push harder, determined to seek entry. Finally past the enforcers I'd fly into a gang of branches that curled around me like fog, only this fog had spines and thorns that scratched me like harpy talons, trying to rip the weapon from my grasp. Some farmers obviously didn't have faith in the sapling hedge so they would run a line of barbed wire through the middle of it, a mere inconvenience compared to the fully grown wooden hydrae that attacked from every angle. To pass through any one of these hedges took over thirty seconds. This doesn't seem a long time but hedge wrestling could steal the energy of the fittest man and once thrown out the other side, I, like many, would fall bedraggled as if literally dragged through a hedge backwards before covering the next person brave enough to wrestle the hedge monster.

We'd pass through countless hedges, each one becoming a little more dislikable - a little like watching the whole of a Eurovision Song Contest. By the end of a patrol we'd punch a hedge if it looked at us the wrong way. If the hedges didn't get us, the bogs or ditches would. A little like Dartmoor, the border areas are covered with peat bogs and ditches for irrigation. Some genius had designed them as an optical illusion - the mind saw a ditch that looked narrow enough to leap across, but once a jump was attempted we knew midair it would be impossible to reach the other side without dismal failure dragging us down into the sludgy shroud of the stream below.

These patrols to dominate the ground confirmed us as walking targets for the hidden enemy. Even covered in bog mud and pricked with thorns, we could easily engage an enemy who would stand and fight, but when emerging like a moray eel to seek prey before dissolving back into the darkness, it made for tense patrolling. While we knew the hidden enemy would strike the first blow there, tattooed within our psyche, was the knowledge that should we make a mistake, we could have our lives ripped apart by lawyers, keen to win the pissing contest regardless of ethics, who would, for months, pore over any soldier's split second decision. Even now, I struggle to find another occupation where an 18 year old's heartbeat judgment may preserve life or cast death and influence a nation's political will.

The hope that our tour would quieten down wasn't helped when we watched events elsewhere in the Province unfold. Three PIRA members killed in Gibraltar saw their subsequent funeral interrupted by Michael Stone, wearing an unflattering flat cap of lunacy, during his killing spree of Republican mourners at Milltown cemetery. Three days later at the funeral of one of Stone's victims, two members of the Royal Signals drove into the funeral cortege in Andersonstown. We watched the sickening scenes on real time TV as mourners swarmed the car like ants around a dead insect, dragging the two men to Penny Lane, a barren piece of waste ground, where they were savagely beaten, stripped, stabbed in the back and shot by people who masqueraded as Christians.

Whatever trouble was happening elsewhere, in the 40 Commando TAOR (Tactical Area of Responsibility) it seemed PIRA had been watching the film 'A Fish called Wanda' and they'd taken on the role of Michael Palin's character, Kevin in their many failed attempts to kill the old lady, played by us. Already they'd lost two on our first day due to bad timekeeping, yet as proof of their tenacity throughout the tour we found plenty of bomb making equipment primed for later use, and when IEDs were detonated, it was a sniffer dog and its army handler that were killed. The death of the handler Corporal Hayes and his dog 'Ben' personally hit hard. We'd escorted them on a route clearance patrol only two days before.

Even when the local PIRA brigade members shot down an army Lynx helicopter flying from Bessbrook to Crossmaglen, they failed to kill a marine. The DShK heavy machine gun they fired managed to shoot away the tail section but due to the extraordinary skills of the Navy pilot he'd managed to bring the helicopter down safely. The only Royal Marine on board defined bad luck - his wish for a smooth flight scuppered as he was returning from hospital after having his haemorrhoids cauterised.

Our multiple was waiting on the helipad to go to Newry when the Lynx was taken down, so as our helo approached we were redirected over Crosslieve Mountain to the Silverbridge area where the Lynx lay like a dead horse. The fact that we were in a Lynx flying into the airspace that only five minutes before had seen another shot down didn't even cross my mind until after we'd landed. Empty brass indicated the firing point so became a crime scene that we cordoned off to await the arrival of the Scene of Crime Officers (SOCO). There we waited, and waited and waited.

'At least it's not raining,' I said to Nick after about an hour of silence.

I spoke too soon.

Various agencies arrived and disappeared, but we remained. Some Royal Marine sweatshirts display the words 'First in Last out' and we were certainly adhering to that. The attackers had gone; probably back home to watch Coronation Street and bicker with the wife that their tea was cold. *Well you shouldn't have take so long in the shower ya eedjit. And why are you burning your clothes? That jumper was a present from me mammy.'*

No secondary device was located nor detonated - it would have surely gone off within the 15 wet hours we laid there. My first real moment of potential action and my role was to lie down on a slightly uneven bank and watch a swaying tree line through the incessant rain.

***

If we felt as if we were just there to be shot at or blown up or get trench foot, our audacious CO had other ideas. A man of many eccentric ideas, his innovative style and love of anything unorthodox made him someone born in the wrong generation yet a perfect leader in such an unconventional conflict. He hatched a plan so devious it could be dressed as Dick Dastardly.

Acorns were planted through gossip spread around those bars in London where IRA sympathisers frequented. Once seeded, back in the Province, Les dropped a military blue mail letter on a road near Dromintee's main housing estate. Written by a marine, the letter was to his mother describing the boredom while on his own at the R21 OP, and the unfairness of it always being him guarding the place while everyone else returned to Forkhill base for their Sunday night shower. Picked up by a passerby the letter contained enough information for PIRA to confirm mainland gossip and to launch an attack on the apparently vulnerable OP. One man wouldn't stand a chance against an Active Service Unit (ASU), who would gain immense credibility by destroying the OP and stealing all the armaments and intelligence gathered at the location.

Operation 'Sneaky Bastard' had begun. I'd wondered why, whenever I was at R21 since virtually the start of the tour, at around 6pm every Sunday, a Wessex helicopter would land for about two minutes, not load or unload, then fly back towards Forkhill. I also wondered why we'd built a waist high screen early in the deployment along the walkway from the OP accommodation to the helipad, it seemed to have no purpose other than to be a makeshift tennis net for a location that had no tennis racquets. As the first ambush personnel of Operation Sneaky Bastard arrived, all became clear. The hope was that a Republican local would pick up the letter, attracted by the fact it was a military blue 'maily'. Of course, the information inside was false but in the hands of PIRA it could encourage them to launch an attack on the OP thinking there was only one man standing guard. In reality there were thirty commandos laid in wait ready to spring an ambush.

The ruse was simple enough. Any PIRA reconnaissance for an attack would see the OP team above the screen walking from the accommodation and seemingly getting into the Wessex that had arrived as usual at 6pm. Unseen to those same eyes, the helicopter would offload thirty heavily armed men who would crawl to the accommodation hidden by the screen. The OP team that seemingly boarded the Wessex would also crawl back to the accommodation to carry on their normal surveillance. The Wessex would then fly off. At this point it appeared the OP was empty other than a sole guard. The Wessex would return a couple of hours later seemingly to return the OP team who would be laying in wait on the helipad hidden by the screen to then walk back in clear view apparently freshly showered. Those on the OP teams even wet their hair before walking back to add authenticity to the operation. Week after week we launched the operation. Some weeks I'd be on the OP team, others I'd be waiting in ambush, hoping that an ASU had abstained from Sunday drinking at the GAA club to take us on.

This audacious operation showed true commando spirit but was too cunning even for cunning paramilitaries - PIRA never took the bait. While it never crossed my young mind at the time, I now look back and wonder how the general public and lawmakers would have reacted to the *agent provocateur* nature of the operation. While the mainland link suggested that approval had been gained from the highest levels, of more concern was how the lads on the ground would have been supported in the aftermath.

A measure of how successful a tour has been is usually surmised by the amounts of incidents while there. On our patch we'd discovered many finds of explosives and armaments readied for disseminating death, yet encountered little in the way of attacks, a true indication that we'd dominated the ground as planned.

Letting our guards down as the final weeks approached was not in our nature. Indeed, with the end so near our concentration and dedication improved to ensure we didn't lose anyone. Despite many of us loving the tour, some of the lads thought I may be tempting fate when I volunteered to stay for another week, to assist in the handover to the 'Maroon Machine' of 1st Battalion Parachute Regiment.

There is much said about the rivalry between the paras and the marines. We have both served around the globe with distinction sharing the workload in all conflicts since WWII. Both elites, pride is intrinsic to the passion to protect our own and usurp the other. There are many men who have failed the commando course yet passed the airborne equivalent 'P' Company and vice versa, and mulish bias will always favour the colour of the beret worn. It's not an argument I readily subscribe to but, if pushed, I'd say if you had an IQ of over 90 and essence, become a bootneck.

On the rare occasion I worked alongside paras, I found them highly professional and wonderfully foul mouthed. The 'who is better' debate isn't supported by any para or marine with over ten minutes service and should be left to the many armchair generals sat behind the parapet of their computer screens, eager to pigeon hole the attributes of each. Different animals dressed in similar fur, we are asked to do the same shitty job.

Handing over to them was an easy task. They wanted to inwardly digest as much information as possible, and were eager to outdo what we'd achieved. It was probably harder for them, as they had to wear our green berets when on patrol to deflect any obvious signs of a handover. Placing a green beret on a para's heads must be akin to placing a crucifix on a vampire.

Our last patrol finished just south of Newry. It would be one to remember. After the Lynx had been shot down in Silverbridge, aircraft would fly at a lower altitude making it more difficult to take on as a target. The bootneck pilot was happy to show us how low he could go. We flew so low I could smell the grass. He bobbed and weaved like a lightweight boxer over hedgerows, under pylon lines and over houses. We flew high, then auto rotated allowing the Lynx to drop like a stone until flying off again high into the air only to auto rotate again. It was the ultimate roller coaster ride, and as our stomachs twisted and turned in time with the roll of the helo, the opposing units exchanged glances and insouciant smiles as if on a Wallace Arnold coach journey. No one would even dare admit feeling nauseous. Upon landing, we ran bent as always to the edge of the helipad wearing the casual look of ambivalence to our breakneck journey. We then went our separate ways to the accommodation and finally allowed ourselves to vomit.

The tour was finally over. Despite the best efforts of the most efficient paramilitary force on the planet, we left the province with the same amount of men as we started - a testament to the men of 40 Commando. It would have been easy to suggest that the threat was over-exaggerated, yet within a week of us leaving, 1 Para tragically lost their first soldier.

There was no ticker tape welcome on our return. We didn't want one. Our satisfaction came from within, knowing the job had been done to the best of our abilities.

I had the rest of my life to ingrain myself further into this life for which I'd given my soul. I was still 18 and had spent the majority of my adult life patrolling the bandit country of South Armagh.

For now, I was satisfied with the six weeks leave we'd been graciously given. I would spend it with family. Not the woman who had given birth to me, or relatives who no longer would recognise the person I'd become; but my brothers with whom I was now irreversibly bonded.

> *'You cannot rouse the animal in man,*
> *then expect it to be put aside at will.'*

> ~ *British Army General WW1*

AFTER AN OPERATIONAL TOUR, it is usually the norm for the unit to undertake an extended period of leave, to reflect on a job well done, to ponder on whether indeed you have made the world a safer place, then drown in the sweet custard of debauchery. Many family men use this leave as a period of quality time with loved ones to see how their children have grown, and to do the jobs around the house that the wife has been totting up over the previous few months. The odd married man takes this opportunity of domestic servitude to steal away with the single lads under some trumped up pretense of being sent on a course - it's amazing how many compulsory parachuting course are held in Mediterranean beach resorts…

Despite living in each other's pockets for the previous six months, invariably many seek to further their connection by holidaying together emphasising the military culture of *espirit de corps*; which in its own unique way is akin to living within a 'family'.

Such was the case after the South Armagh tour. Eight of us descended to the popular Greek Island of Corfu at the slightly less-than-classy resort of Ipsos. Famed in those days for its rowdy nightlife, cheap beer and even cheaper morals it seemed a perfect destination. As it turned out, we weren't the only group from 40 Commando to descend upon its fag-butt infested pebbled shores - At the final count thirty five bootnecks had arrived - more than troop strength.

This was my first holiday without my parents, so was acutely aware of not bringing my teddy and making sure I wore clean underwear daily. The latter was not overly concerning as I'd found the advantages of 'going commando'. Not only was it more comfortable, but also it had the advantage of less laundry and less money spent on clothes meant more for carousing. The downside, of course, would be skid marks in my trousers and any unforeseen dribbles would be highly visible, especially in trendsetting cream chinos. For some equally strange reason wearing of lycra shorts also became fashionable. Many lads in the Corps - including me - decided that they were so en vogue they should be worn not just for running or cycling, as was their purpose; but also at every other opportunity. I can let your imagination work out how someone going commando would look in skintight lycra shorts walking down to the corner shop for a pint of milk, or as we did - worn in the crowded bars of Ipsos. It's not pleasant - a three-dimensional map of male genitals, forfeiting the privacy that wearing normal shorts would afford. It may, however, remind you to buy some cocktail sausages or prunes.

Being away for so long without even the smell of perfume to waft across eager nostrils, it is easy to see why a young red blooded male would seek female company when returned to society, especially with my newly found skills, I was feeling quite benevolent in my sexual pleasure. Rather than just concentrating on my own selfish endeavor, I could now become the Salvation Army of sex, the Oxfam of oral, yet hopefully not the gift aid of gonorrhea. Yet despite the intoxicating mix of sun, sea and sand; my semen-laden donations didn't quite have the receptive audience I'd hoped. The lads, on the other hand, had far greater success with the sunburnt ladies in lycra boob tubes, more so if they were on the lower end of the desirable scale who were bagged off in their droves as bootnecks are true trail-blazers of egalitarianism, shagging anything, irrelevant of race, creed, religion and in some cases gender; and followed the Moroccan proverb that 'every beetle is a gazelle to its mother.'

The popularity of Ipsos meant that I bumped into some old school mates who were also holidaying there. What struck me was how skinny and pale they looked compared to us, even if we were dressed in varying themes of silly rig.

Nick's adornment had gained the most favourable reviews after deciding to while away the evening dressed as his alter ego superhero 'Lilo man', resplendent in a pair of 70's swimming trunks bought from the widow's auction of a deceased Marine, a white T-shirt displaying his superhero's name and a bright pink un-inflated lilo as a makeshift cape.

We met up with four other newly arrived boot-necks, including Sandy, who'd reached legendary status on a previous deployment to South East Asia for winning a Muay Thai boxing bout with a broken arm. Will, the only non ex-boxer in their group, was absent as he'd decided to stay alone in a Welsh bar to try his luck with a girl.

Will eventually joined us, his head mutilated, his face covered in blood. No, he hadn't been committing cunnilingus on a menstruating drunk, although he wouldn't have been the first one on the trip. He explained that he'd received a rather unpleasant kicking from three Welsh guys. It would have been easy to subscribe to the notion that 'the bigger man walks away' but the DNA of many of the lads meant that passive resistance to being attacked was never an option - we abided by the motto 'never start a fight, just finish it'. Six months of frustration fighting a hidden enemy, six months of having our hands tied behind our backs by politicians withholding the inner beast, we were suddenly catapulted into a situation where the gloves were off. No one could tell us what we should do, we already knew. No longer were we in the dirty war of the Northern Ireland conflict. We were now at the Battle of Corfu.

Without any form of hubris, I would say picking on a bootneck is generally a bad idea even if we were ridiculously dressed. Baz, a corporal from my troop, called what could only be described as an 'O' group at the bar. Orders were briefly given as to the plan of action - basically up the testosterone levels, go to the Welsh bar and kick the shit out of Will's assailants and anyone else who wanted some action. I felt a familiar hype burgeoning inside, my heart was beating hard underneath my balloon breasts and obscene t-shirt, accelerating as I speed marched down to the Welsh bar together with more than a dozen wig-wearing boot-necks; ra-ra skirts fluttering in the sea breeze and the pearlescent reflection of gold and silver lame sparkling under neon streetlights.

The Welsh bar sat at the bottom of the club strip. We turned into view of the bar, looking like a burly drag queen dance troupe, the steely stares through various shades of eye shadow - 'vivid blush' being quite popular - told of the impending conflict, compelling the bar's clientele to seek sanctuary indoors.

One lad stood bravely outside the bar, his arm bar-ring entry from the doorway. 'You're not coming in lads.'

Baz, the spearhead of our group and probably the most daring of all in his low cut chiffon dress, threw the lad by his shoulders into a table. Continuing through the doors, straightening his wig, for it just wasn't right to fight with errant hair, he waved for us to follow. We followed on excitedly into the dark interior unaware of what would meet us on the blind side of the door.

Inside, hell was unleashed. Bootnecks, high on revenge and nail polish fumes, just hammered their way into every man in their way. Fists and oddly angled bodies flew everywhere as we pushed through the dimly lit bar. Amazingly, I hadn't yet been hit but had managed to break a false nail. Nick, who by his own admission wasn't a fighter especially when his lilo cape had been wrapped around his face, had managed to hurt somebody with a bar stool and was by my side as we ran the gauntlet through to the rear of the bar. Optics behind the bar were being smashed to my left, faces smashed to my right. I'd managed to get to the rear of the bar without even swinging my handbag. Now isolated, I was glad to hear the Baz's order of, 'Re-org!!' We clumped together and backed out of the now decimated bar into the cooler night air to meet the sound of glass underfoot, chaos of panicked tourists, and of an approaching coach onto which one of the fighting Welshmen jumped. In my moment of vainglorious bravado I followed him onto the stationary coach. I stood in the doorway like a cheap hussy advertising sexual favours and offered, 'You punch me five times and if I'm still standing, I'll punch you back…err five times?' - A sentence that was a top contender to win 'The Most Stupid Sentence of the Year' award as voted for by the readers of 'What A Dickhead' weekly.

It seemed like a fair deal. He took up the offer. While it seemed his fighting skills had been based on some 'Keystone Cops' movie, his repeated punching of my idiotic head left the iron taste of blood swilling around my mouth. It did occur to me whether it was such a good offer after all.

'You finished?' I asked, trying to sound as cool as one possibly could with a slightly insecure tooth, a battered face and a ripped 'I Love Cock' T-shirt.

He looked a little surprised. It was clear he had hit me five times so in the name of fairness it was an irrelevant question.

'So it's my turn?' Again, I can't have looked particularly threatening with my pink lycra mini skirt riding up to show my stripy g-string.

He didn't respond but with guile as plentiful as his counting ability pushed me off the entry step into the street. Attempting to re-board the bus became futile. I jumped straight into the closing doors trapping my bicep in the process, leaving me with bare arse cheeks glowing under the bar lights looking like a cock-loving vet with my arm up a mechanical cow's arse.

Will was in a bad way, so was bundled into a taxi to transport him to hospital with his close mate Screwy, whose massive biceps looking engorged within the luminous pink tightness of a size 12 lycra number. As police sirens approached, we decided hanging about wasn't the best idea so scarpered as fast as we could in crippling ladies footwear. Thinking we may be followed for a revenge attack, we decided on splitting up to meet at an emergency rendezvous point (ERV) where we had a vantage point to our apartment.

It was hardly the escape from Sobibor, but only five of us made it to the ERV. We waited silently until confident that returning to our first floor apartment was safe. To be doubly sure, once inside we barricaded ourselves in, pushing every bit of heavy furniture against the doors and French windows. We even thought of tying bed sheets together should we need to make further escape over the rear balcony.

Laughing in the afterglow of adrenaline, we sat looking like transsexual rugby players covered in dried blood, smeared lipstick and disheveled wigs. Bryn had turned his ankle for which he deservedly received shit - we'd told him previously 6" heels were so 1970s.

Our revelry was short lived. A crash of the front bedroom window shocked us. Our room was purposely darkened to give us the ability to look out through the window nets without being spotted ourselves. Outside stood a group of about forty men armed with chains and bats. They were clearly not cricketers or late night jewelry salesmen. Our plan of not being followed had evidently gone a little awry. We'd just returned from a civil war zone and now in the peaceful surroundings of the Greek Islands we were closer to death than ever.

The mob's heavy handedness played into our hands. The frightened owner of the apartment locked the front entrance offering us an extra layer of physical defence. However, their shouts, threats and general scaremongering did little to quell our mood. Combined with our alcohol intake, adrenaline and energy levels were now falling, so in the best traditions of the military we set up a sentry system where one person would survey our hostile surroundings while the others dozed off to alcohol assisted slumber.

The loud bang on the apartment door startled us into a state of blurred alertness. We looked outside into the dull light of a sepia morning. The gang had disap-peared and only a disinterested goat with an amazingly large udder trotted the hotel path. Had the gang outside managed to overpower the owner and surge up the stairwell to strike, with great vengeance, motorcycle parts and sporting equipment upon our heads? Was it the police, looking to arrest us all and send us to some decrepit forgotten jail to introduce us to sweaty and overly hairy men who had forgotten the touch of a woman, so instead made merriment with young men, handcuffed and handsome? No, it was Will and Screwy. With their eyes blackened and closed, it was a wonder they'd managed to find their way home. It seemed they'd been repeatedly beaten with something far harder than a mahogany baseball bat. Screwy's head was the size of a well-tended pumpkin, although, in truth, no one noticed the difference. Will had looked pretty bad after his first kicking but looked as though he'd been in a fight with a blood transfusion bag. He sported a comedy nose twice its normal size, which made it four times the size of an average nose. They enlightened us with their tale: Arriving at the hospital, they'd been seated near some of the Welsh bar crowd also receiving treatment. Reporting Screwy and Will to the police they were dragged from the haven of a nurse's gentle touch to the less attractive grasp of the police at the local station where they received some interrogation techniques that weren't on the UN approved list. They'd been there six hours and not even been offered so much as a biscuit. Instead, they'd been handcuffed and their heads used like rubber balls against the wall. This only galvanised Screwy and Will's resolve, neither uttering a word to explain their actions; either that or they just didn't understand Greek. The police chief conducted summary justice by insisting they'd have to pay for the bar damage. Chaperoned by the police back to their room, after handing over all their cash, they now had nothing. Not a penny. The pot they had to piss in was now amiss. With a week left on the island, the daily jetski rides would now have to be forgone, and eating would be optional.

Being close to the end of the holidays funds were low, but digging deep we gave our share of the fine to compensate Screwy and Will. We weren't sure whether these fines filled the back pockets of the bar owners or the police, but certainty was in the cleaner's mind when she outlined our safety was in jeopardy as the Welsh bar belonged to the local Greek Mafioso. We were in shit so deep we'd need a submarine to escape. Having the ability to fight people bare knuckled was an honourable, if not outdated, skill but wasn't particularly useful against chainsaws and although we were all immensely strong swimmers, concrete boots tended to restrict even the most competent. Our cleaner continued the good news by reporting gun toting gang members at the bottom of the road waiting to exact revenge. This lightened the mood considerably; we all agreed being shot was preferable over torture.

Being hunted by the mafia put a bit of a downer on the day. We ventured as far as the bar below to eat more pizza and counter-surveilled strange looking men waiting out on the road outside observing our mastication of stringy cheesed savouries. We knew we were good looking but were they mafia, cheese fetishists, or were we just paranoid? The argument was irrelevant. We were clearly no longer welcome in the resort. Having only two days remaining in Ipsos, we decided that we'd outstayed our welcome. So, like any self respecting commando would do after a successful raid, we planned an early morning escape, evading any mafia hunter force, and headed for Corfu town.

Corfu town in the height of summer will always be pretty bereft of hotels that willingly accommodate eight men sporting various facial injuries for just the one night. Fortunately, there was one hotelier whose accommodation was so awful that they couldn't even attract tourists in the peak season. The place would have made the local prison look like some 5 star

Maldivian resort. Our new home for the last 24 hours of the holiday, one that we had built up so much and excitedly pored over the previous six months, ended in a hotel corridor shared with a host of dishevelled down and outs who clearly enjoyed angry monologues and projecting sputum. At least the corridor floor had mattresses, even if they did come with a free dose of tuberculosis.

***

Post leave, the drafting desks of the Royal Marines had to do their bit of manpower Ker-Plunk, leaving holes in less important places while keeping numbers high for those with imminent operational deployments. Having just finished ours, the men of 40 Commando were scattered to the four winds. We did have an influx of new lads coming straight from training, all bright eyed and bushy tailed, looking for guidance from the senior guys recently returned from Northern Ireland. It'd be easy to cast off this new breed as faceless sprogs. It was a given that they'd get a certain amount of shit as new joiners and the gullibility average would be heightened during grot parties; but under their wiry bodies taught from the traumas of commando training, recruit hair and spotty skin scarred from Woodbury rash, potential abounded - young minds keen to absorb alcohol, information and new skills, jumping in with gusto to each task given to further themselves as not only commandos but rounded, rather eccentric socialites - and nurturing that potential was key to put them on the right path.

There is little more satisfying during a military career when you look back and see, even amongst this tiny group, their most rapacious cider drinker becoming a national safety manager for offshore oil platforms, my personal NAAFI runner eventually becoming the RSM of the Special Boat Service - amazing really when he returned one day with a Snicker when I'd clearly asked for a Double Decker - and one fellow Yorkshireman who wouldn't let the handicap of being 2'6" with a squeaky voice prevent him going on to gain an SF commission after being immersed in that incomprehensible netherworld of strategic clandestine operations where Jason Bourne's testicles would shrivel in fear. All great people have begun their journeys somewhere and there is arguably no better place than the Royal Marines.

Having some seniority allowed me to get onto a few short courses around the country and becoming a service rifle coach allowed me to get some instructional experience under my belt. I gained a superior pass, surprising considering I was the course sprog.

One of the fellow students, Geoff, wouldn't get a superior pass. In fact he wouldn't get a pass at all. He'd get RTU'd (returned to unit). Geoff liked a beer. That's wrong, he liked many beers. Pissed as a parrot after his Sunday lunchtime session at home, he boarded the Derby to Exeter train to get to Commando Training Centre where the course was being held. After a few more warm tinnies on the train that were always useful as piss receptacles, he fell asleep. Not hearing the train announcement to alight at Exeter St Davids, he woke up at around 10pm with a tongue like a Tasmanian leper's Odour Eater when the train stopped in Plymouth. Never a man to panic, he went to the nearest pub to await the return train back up for the hour and a bit it would take to get back to Exeter. It was late by the time he set off again and despite him having a drunken nap, his body decided to enter into another deep sleep. He woke up just in time. Just in time to alight at Derby his original boarding point. He managed to arrive back at CTC by Monday afternoon after spending around 21 hours in various degrees of slumber on a train.

Geoff's antics though paled into insignificance when compared to the only sergeant on the course - Ed. Renowned for wearing a baby doll nightie under his working uniform, he was happy to show anyone interested that he could self pierce any part of his body. On a Wednesday night in a quiet Exmouth pub you don't expect to see a sober man stab a kebab skewer through both cheeks, use a safety pin to pinch together both lips and his tongue or pin his foreskin to his belly.

The course itself had taught me many skills other than how to self mutilate one's face and penis. My instructional technique had reached new heights and so too my shooting to such a degree I was invited to join the Corps shooting team. Knowing there were some decent hot trips coming up with Bravo Company, I turned the chance down of spending two months shooting in the United States. I immediately regretted it.

4 Troop had a new Sergeant. Scouse was replaced by Stan - a man who made me look tall. His pugnaciousness totally changed the dynamic of the troop.

Many new to an appointment will rightly stamp their own individual brand of authority, yet his style was perceived as one of continual contempt and destroyed Scouse's legacy of harmony.

As someone who'd failed to grow to any notable height, to describe someone as having a 'small man's attitude', does not often sit well with me. However, I could see why others in the troop described his character as such. I satisfied myself with the un-heightist view that even if he was 6'9" I would still voice the same opinion of him (but do it far quieter).

He reveled in the dominion he now held over his new charges. Rather than implementing any form of edification to his young charges, his encouragement was usually a brief, 'Sort it out lofty, or I'll fill you in.' His obvious intelligence belied his mantra that toughness escalated with rank, easy to do when hiding behind the immutable military legal protection of hierarchal power. Unfortunately, this theory is easily shot down by saying the Captain General of the Royal Marines is Prince Phillip, and I'll wager most marines could be victorious over him in a naked jelly wrestle.

Still a young marine, crowbarring my sometimes unnecessary penchant for loquaciousness into any debate, meant there was only going to be one winner in a verbal joust with any SNCO, never mind Stan, so I tried my best to steer clear, yet it was difficult considering his omnipresence around the grots.

Going into Christmas leave, there would be no separation anxiety from Stan, but on the final morning, he called us together to seek volunteers to fill vacancies for forthcoming specialist courses.

The signals course was dreaded by most. Most on the course were treated terribly and Signals Troop was hardly one to aspire to. While they had some fantastic, switched on lads, to be press ganged to join them wasn't favoured by many.

No one raised a hand to volunteer. That was corps tradition. Stan then asked us to raise our hands.

'OK, put your hand down if you have been in the troop less than six months,' said Stan. The vast majority lowered their hands.

'Lower your hands those who have a draft coming soon.' Many hands again were lowered.

'Right, who already has got a specialist qualification or technical qualification?' A few more withdrew their arms.

'Who has a course coming up?' Again a small number lowered their mitts.

I looked around. There was two of us with arms still raised.

With a glint in his eye, Stan told us, 'Well volunteered. You're on the next sigs course.'

Trying to forget what shitty future lay ahead of me, my mind turned to the skiing holiday in Tignes planned with Nick, and two other guys from the company, Popeye, and Harty. All but I had skied before, so I felt guilty as Harty wasted his first day on the slopes trying to teach me how to stay on my feet. He was as frustrated as me. There was clearly a gaping chasm between theoretical and practical attainment - stem christies that Harty made look easy became just the continuous aggravation of cross-legged trips where my body would fling itself head first into the nearest lump of hard ice, and trying to dig in my edges just allowed me to fall over sideways. My impatience of not being able to do something immediately, especially sport, was becoming more pronounced. I found skiing something I wasn't particularly a natural at, so I slumped from the first day's slopes pretty pissed off that I wouldn't be an immediate winter Olympian. Even dressing up as a cheerleader and getting so drunk I walked five miles in -15 degree temperatures claiming that there was a frilly cardigan in every drink, couldn't hide the fact I was alien to this snowy environment and a complete ski biff.

I decided to spend the next day on my own to try and learn through self-instruction and to give the others time to enjoy their well deserved break. Trying to recite what Harty had taught and squeezing a few pips of advice from Nick, I began to stem christie quite adequately. Admittedly, I didn't quite have the grace of an equine student of the Spanish Riding School, more of an ostrich with vertigo. Nonetheless, I hit the button lifts with the competent skiers, and realised I wasn't yet a peer. I made a complete arse of myself. Repeatedly. Thinking I'd now overcome the complexities of sitting on a lift I ended up falling off again halfway up the hill. Attempting to traverse back down, I unfortunately couldn't avoid the nearby slope that was being used for a downhill ski competition. With professional skiers hurtling past, there was I plodding across the fall line like a moving slalom gate, trying my best not to get killed by a hurtling skier or an irate official trying to chase me down. I must have looked like the world's most overdressed streaker.

By day three I was ready to take on the 'egg' position. A bootneck favourite, the 'egg' is how you see a downhill skier fly down the slopes. Point your skis downhill, crouch and just go as fast as you can, bouncing men, women, and children out of your way. I took to egging pretty well. It took little skill, just balls; the only technique necessary was the ability to stop without the aid of an immovable object.

Tignes has a huge ski field, combined with the neighbouring resort of Val d'Isere there are over 300km of ski runs to break your neck on. I was having a ball and quite happy to go off on my own to explore the more remote slopes as if some seasoned off-piste extreme skier.

There is little more enlivening than the feel of ice cold air masking your face as you hurtle down a smooth dry snow slope with no one to get in your way. I couldn't have felt better as I hurtled down an exceptionally steep, yet smooth, empty red slope. I can't have been alone, an errant ski passed in front of me. It seemed familiar. It looked the same one as my left one.

Of course it did. It was my right one. I was skiing on one ski. My rear binding had just decided to fall off leaving me skiing on one leg. It seemed an eternity but must have taken a millisecond for me to crash into the slope and roll heavily down the ice taking away a bit of facial skin and lots of pride. I recovered my ski, now of little use minus a binding. What was I to do? It was quite a way down to the ski lifts, so carrying my poles and a useless ski, I was forced into a rather ungainly one legged ski technique. I lost count of how many falls I suffered after I lost the will to swear any more.

Managing to get to the lift, by pure chance a maintenance man was doing some rudimentary repairs, so in my best attempt at French asked, 'Avez vous le harry black maskers?'

With some neutral hand signals we managed to understand enough to solve the problem with his electrical tape strapping my boot to the binding-less ski. I was about as far away from my hotel as possible without actually being in another time zone, so the next three hours was spent trudging, from ski lift to ski lift. Button lifts were impossible to use, as the tape prevent-ed the ski from sliding on the snow, so I was basically bottle necked into using chair lifts as a hand rail back to my resort, skiing pathetically as best I could on one ski the other limply hanging in the air like a dog with a thorn in its paw.

I thanked my lucky stars that I was a sunshine commando.

'America, fuck yeah.'

~ 'Team America'

MY LIFE AS A GENERAL DUTIES MARINE also seemed to have taken a nosedive. With a signals course in the offing, I had to find a way to get out of it. Claiming I was gay wouldn't help - many familiar signalers were the happiest to engage in many a homoerotic challenge.

Fortunately on Daily Routine Orders, one marine was trying to get out of an Assault Engineer's (AE) course. Who wouldn't be happy to learn how to use explosives, set booby traps and drive small inflatable boats? So, in a moment of impulsive desperation, I took his draft. Within three weeks I would be back at Commando Training Centre to commence my AE 3's course, together with Dinger.

I knew Dinger as a child. A year younger than he, we went to the same secondary school, and while we didn't hang around in the same group we were involved in some football vandalism together as young pseudo hooligans following Leeds United. At an away game at Huddersfield, I remember being impressed by his accuracy when throwing a number of house bricks at riot police. He joined the Royal Marines a year before me, therefore had taken an interest in my career and we talked occasionally through my early years in the Corps. We now became closer as students on the same course and together would join AE troop back at 40 Commando once finished. Immediately, healthy competition brewed between us. I would come top of one exam, he top of another. His practical work was immaculate so I used him as a benchmark, trying to outdo him wherever

I could. He had the brains and aptitude to take each task on its merit. I was glazed with creativity using flair to achieve my aim, sometimes with less than satisfactory results. We were tasked to booby trap certain items and I managed to booby trap a Mars Bar, fitting a miniature pressure pad inside that when activated by a bite would take your face clean off, which was handy should you wish to assassinate a fat politician, obese from years of sticking his nose in the trough. The construction side of things was less pleasurable. I got little stimulation learning how to build a bridge, pour concrete or raise scaffolding, all quite useful skills in civilian life, but I was a bootneck, and civvy street was as alien as Jupiter.

I placed second on the course, attaining a distinguished pass, which was quite a high accolade; one I totally took for granted. I hadn't really wanted to go on the course, volunteering only as it was the lesser of two evils. Dinger obviously came top. His application and strong work ethic came just behind his integrity; characteristics, that not only sculpted a fantastic human being, but would stand him in good stead throughout his career. Dinger was clever enough to know his weaknesses and would work hard to improve in areas he needed to. With grit, determination, and a high degree of competency, Dinger would become a helicopter pilot, reach the rank of Captain and become a high flyer in civilian life. Not bad for a kid who was brought up on one of the roughest council estates in West Yorkshire, and threw bricks at coppers.

Life in AE Troop of Support Company couldn't have been easier. We did physical exercise, checked equipment, played darts, drank tea and fixed the CSM's cistern when broken. It wasn't what I had in mind for a life in the commandos but it was better than being a signaler, and my ability to hit a double top had greatly improved. With only eleven of us in the troop, we varied widely in experience and age, an odd mix that knew each other well but not close enough to know each other's dark secrets.

What we did know was that our Troop Officer, Colour Sergeant 'J', was known as Colours 'Drunk,' yet he commanded respect and knew his onions. He was pretty laid back and as long as the jobs got done he was a happy chappy. I too was an exceedingly happy chappy when he told me I was going to go on the next Mediterranean trip then straight to the West Indies for four months of sun, sea and safety fuse.

We boarded the Royal Fleet Auxilliary Landing Ship Logistic (LSL) Sir Geraint sailing in a fleet with its sister ship, the Sir Galahad II and HMS Intrepid. LSLs are flat-bottomed giant surfboards that cut through the water like a chipped house brick. Even a small ripple on the quayside can cause one to sway like a drunk. Therefore, as we approached the Bay of Biscay, infamous for its swells, I began to feel slightly seasick. The sickness worsened and from feeling as fit as a fiddle, I excused myself from lectures to lay on my bunk feeling hugely sorry for myself, not eating and trying not to vomit every time I became vertical. Scran times became ever more infrequent and my final attempt at sustenance became a fruitless attempt at trying to prod a rolling sausage on the sliding metal plate that belonged to the guy sat next to me. The plates, compartmentalised for each course, became a messy mixture of each dish once the ship rolled. Thick gravy slopped into rice pudding that would then inundate the minestrone soup - an edible Dali painting for those with the hardiest of constitutions.

Chinese whispers dictated that an LSL could tip over at 45°. We'd already experienced a 43° list, and as the Bay of Biscay rocked and rolled like Elvis on a bender, I deteriorated further into a sea-sickened human blancmange.

Seasickness brings with it a special form of misery, where solace only comes in the form of death. I puked myself inside out, my duodenum the only organ left to yet pass my lips. I couldn't stand without puking. I couldn't lay down without thinking I was to puke. Some amphibious commando I was. As the emergency drills were practiced again, I ran to the swing door toilets only to get stuck as my flotation device was too wide for the cubicle. There I stood as pathetic as a man can be, stuck in a narrow cubicle doorway, smelling shit and dribbling translucent bile down my fluorescent orange lifesaver.

I had to find some form of cure, so was dispatched to the sickbay by those with a far better constitution than me.

On an LSL, the sickbay is on one of the upper decks. This means that any ship roll is amplified. The higher I went, the sicker I felt. I wasn't sure I would make it without falling into some starvation-induced vertigo and fall down the many steps I'd climbed. While waiting with a sick bag in hand, the Doc explained that a jab would probably be the best option. A pathetic wan nod and wretch told him that I would welcome anything, including a firing squad, to relieve the anguish.

What wasn't as welcome was the MA who was to jab me. He wore spectacles that I would, in my childhood, have described as 'jam jar bottoms.' Here I was, bent over with my pasty arse in the air exposed to a myopic MA, squinting and stumbling with the violent rock of the ship holding a needle that was far longer than I'd hoped. Feeling like a moving dartboard, he luckily managed to get the bull's eye and hit the juicy part of my nubile, firm, sensuous right buttock.

The ignominy of receiving a jab paled into comparison of the relief it brought. While I wasn't 100% I could at least eat small amounts and the scurvy I thought I would suffer never materialised.

***

Dull would be the soul of any Royal Marine who could ignore a sight so splendid in its Majesty as the Rock of Gibraltar. The Royal Marines battle honours are so vast it has a globe in its emblem to tribute their success in every quarter of the world. Yet above the wreathed globe sits the words 'Gibraltar' betrothed for the Marines' success in the Siege Of Gibraltar in 1704.

To visit Gibraltar as a Royal Marine is a pilgrimage of battles fought wherever they may have been, and the first sighting of Gibraltar rock is a sight to behold. Jutting from the water like a God commanding the straits below, it's easy to see why this monolithic giant is so strategically important for empires to control.

Desperate to set foot on land, the whole unit jumped ship as if it was about to sink when the gangway was set down. An amphibious landing of lithe men wearing T-shirts emblazoned with foreign destinations, Union Jack shorts and desert boots swayed inside sea legs as equilibriums hadn't quite adjusted to terra firma. Seemingly, the only attractions were the pubs and bars that survived on visiting British Military personnel. For Gibraltarians, another injection of money was welcome. Usually the first port of call on any ship journey from the UK means servicemen reach deep into their pockets in search of fun. In 'Gib', for large groups of men, there is little fun to be had outside of the pubs.

Bootnecks and matelots were everywhere in the spring sunshine. Pubs were full of well-humoured banter, and I found love in the shape of Pink Gin; a wholly feminine name for a drink that I managed to scupper countless numbers of. It seemed to be the drink of Gibraltar and those who'd seen my inability to sail now saw me holding a pink drink, while talking to Charlie, the owner of 'The Hole In The Wall' bar, must have wondered how the fuck I was a bootneck, especially when they were more used to the shenanigans going on next to me where six bootnecks were drinking straight from an industrial sized jar of pickled eggs, each consuming one egg at a time, with vinegar, until the jar was emptied. The fact that one of them hurled half digested egg, vinegar and beer back into the jar didn't end the game…

It was FA cup semi-final day back in Blighty. Many of the unit Scousers were in happy mood as their beloved Liverpool were playing Nottingham Forest at Hillsborough, Sheffield. Glee turned to disbelief when news filtered through about the many Liverpool fans crushed to death. Grief turned to anger as idiotic local civilians taunted the lads about the incident, yet unbelievable restraint was shown. Footballing rivalries were forgotten as lads came together as one with their military compatriots and I, for one, would have thrown some of the sick bastards into the harbour, which would be prophetic considering what was to happen. Of far less importance, the NAAFI Manager on Sir Geraint seemed to have gone AWOL, leaving some ranks on board mutinous.

The 'Top Of The Rock Race' was obviously the idea of a teetotal racing snake. 2.7 miles uphill to the 1300ft summit of the rock, it was tradition to run this on a Sunday morning at 0900hrs prompt - about 5 minutes after being dragged from your pit with a mysterious 'Made In Gibraltar' tattoo ringed around the navel or when most lads were returning, bleary eyed, from ashore. The course is exceedingly scenic, should you not be breathing out of your hoop and avoiding the arse aromas of the runners in front. There is forever a football crowd of Barbary macaques watching - I say football crowd as most watch patiently, yet there are a small violent minority who attack anything they see on their 'manor', including slow running bootnecks. A lad known as Rocky lived up to his nickname by instinctively punching an aggressive monkey that attempted to attack him as he rested by a wall. Violence is never the answer - unless a twat of an ape is attacking you. The race certainly clears the head and would be a great hangover cure for those who see the attraction in running up a big fucking hill.

Only after we returned from running the rock did we hear of the NAAFI manager's whereabouts. His body had been found floating face down in the harbor stuck in between two of HMS Intrepid's landing craft His death was never explained to us, yet rumours flew carelessly. It had been an eventful few days on my first overseas trip.

***

Tactically, our first event was a combined exercise with the French, Dutch and the US on the scenic island of Sardinia. Again, my first taste of multinational operations, I was excited about working with other nations. It'd be a while before I got to see them. Painstaking route clearances didn't seem like real commando operations as we pushed through the scrub and dusty tracks, but we were well in advance of the US Marine infantry who would pass through much later. Upon task completion, we sat on our arses for the next 36hours awaiting the US Marines to pass through. Time goes slowly when working with other nationalities.

Returning to the multi national HQ camp on completion of the exercise, it was clear who were the resident vagrants. The US, Dutch and French all had huge orderly tents to accommodate their soldiers. Under the Union flag were the leads of 40 Commando RM, accommodated under bivvies that hung scruffily from any solid object, be it a tree, land rover trailer or another bivvy attached to something remotely solid.

Other nationalities constructed separate recreation areas where murmurs and the sound of a radio would play from a tent. Over in the UK section, around a huge bonfire made from a pile of crates, pallets and used boxes, one could hear, 'Holes in hands, holes in feet, he carries crosses down the street. Has anybody seen JC, JC JC JC…' sung from the three badge marines, holding cans of warm beer, passing traditional corps drinking songs to us sprogs who would in future years pass down their legacy. The dulcet tones of 'Kit Kat she wanted, four fingers she got,' carried away on the burning embers, filling the smoky air that stung the eyes. I hardly knew any of these old blokes with thick moustaches and ruddy complexions but listening to them with hilarity certainly made me feel more engrained within the fabric of the Corps.

The unruliness of our location showed why the British military are so highly regarded around the world. We do better than the rest with less. Shit kit, crap equipment and relatively small numbers teaches resourcefulness, robustness and unconventionality, all necessary traits as a modern day commando. Happily biased, I consider the British military the best pound for pound fighter, plentiful with guile, intelligence and cunning to perfect its craft. I looked over to the Stars and Stripes fluttering above their impressive tented city. The US military - a disciplined heavyweight Queensbury rules champion, with heavy weaponry second to none but with weakness against those smaller, of greater speed, innovation and eccentricity. However, with its immense strength and power one lusty blow from its gargantuan glove the US can knockout any foe no matter what its capabilities. Virtually every US serviceman I ever met was straight down the line, totally committed, focused, yet veered away from the humour that united us bootnecks; their pursed lips hiding Hollywood teeth in stark contrast to the crooked pegs that shone yellow within a bootneck's permanent smile. The American mission was all-important. Any soldier who steps out of the tight indoctrinated guidelines is likely to be cast out. Mavericks in the conventional US military do not prosper, no matter what Tom Cruise tells you.

Over near the beach sat the Dutch contingent of Marines. To a man, tall and thin, every cloggy marine must be stretched on a rack in basic training. They seemed to reflect the stereotypical image of the Dutch. Laid back, chain smoking fun loving characters, but with long hair. Their particular unit was soon to be disbanded so flew their unit flag from the highest flagpole - a beacon of pride, a fluttering salute to their forefathers. Somehow it had become lost and accusation that we'd stolen it were vehemently denied, although it did mysteriously reappear a few months later, relegated from its original salutary purpose to a makeshift bed space divider in one of 40 Commando's grots.

Having bought a book on the French Foreign Legion in my days prior to joining up, I was especially interested in their camp. I loved the romance of the 'Legion Etranger'. A hotpot of loyal Francophiles and adventurous familiaphobics with a generous dash of psychopathic thugs; I often wondered whether I was more suited to their lifestyle than the one I'd chosen. While I couldn't think of anything worse than drill, I still watched slightly in awe as they purveyed a permanent portrayal of their culture through pomp and ceremony. Even just going to their meals was a slow marching operetta of songs celebrating victorious battles or lamenting abject catastrophe - a rather more salubrious affair than watching two baggy arsed Bootnecks, berets tucked in their pockets, talking about the previous night's bag off.

Cross training allows a certain amount of appreciation of other nationalities combined with a heavy dose of posturing. When it came to kit and equipment, the Yanks kick ass. They took us joy riding in their brand new Hummers, allowed us to go into the sea in their huge amphibious AMTRAC vehicles and even took us SPIE rigging, (Special Patrol Insertion and Extraction) a method where you are shackled to a long rope under slung from a sea knight helicopter, then flown away, dangling hundreds of feet in the air, an amazing fairground ride for Special Forces. In return for all their hospitality and generous demonstrations, we showed them our unreliable SA80s.

The piece de resistance for the multi national force was a firepower demonstration by the US. We marched up onto a high cliff overlooking a bay, a natural amphitheatre of destruction and mayhem, eager to witness the world's most powerful military machine doing its thing. It was a very impressive sight should loud bangs and explosions turn you on. The commentator narrated in detail which weapon system fired what ordnance and what carnage could be expected with subliminal messages of how they could kick any of our arses should we get a bit lippy. As more explosions gleefully destroyed the eco structure of the bay the commentator announced the finale would be a B52 bomber from a strike command somewhere in the USA that had flown all the way over especially just to drop a payload. Although I've never been part of the process that goes into dropping bombs, one would imagine there's rather a lot of planning, scheduling and detail to ensure that a B52 can fly 9000km over the Atlantic, across the Iberian peninsula to carpet bomb a target destination on the Island of Sardinia at the end of an unrehearsed firepower demonstration. En route, I'd have liked to witness the professionalism of the aircrew, final checks made, as the destination approached their focus would intensify to ensure the bombing was a success to posture to those watching how good the good old USA was at bombing things, as if we hadn't seen enough examples in the 20th century. I would have loved even more to witness them on their 9000km return journey discussing how, after all the meticulous preparation, they'd just missed their target. Luckily they were too far away to hear the chortles and sarcasm of the British, French and Dutch troops sat on the hill above the target range. It'd not been lost on many that the USAF had just made an 18000km round trip at huge expense to the taxpayer and failed to hit the second largest landmass in the Mediterranean Sea. So much for precision bombing.

***

Phase one over, we cross decked onto HMS Intrepid for a trip across the Atlantic Ocean where we'd spend a couple of weeks in Florida before cruising around the Caribbean Islands on a jaunt masquerading as work.

HMS Intrepid was so old it was commissioned when Lord Nelson was a leading hand, but had served the RN with distinction. It looked a salty sleek grey messenger of death for two nautical miles when it had left Portsmouth Harbour. At that point it reverted to type - a battle ravaged geriatric - its creaking engines catching fire. Luckily with the amazing resourcefulness and endeavour of the sailors on board armed with fire extinguishers, masking tape and paint it'd been fixed before the Admiralty could call her back in and scrap the West Indies tour.

Being a small contingent, AE troop was split into different mess decks. Jammy became my bunk pal in mess 4L2 along with fifteen resident stokers. My bunk was bolted to the engine room bulkhead. The incessant vibration and noise I thought would become a sleep depriving annoyance but after a couple of days it became a mechanical lullaby humming me to sleep easily after a couple of sociable CSB beers. Another Bootneck, Dolly, was also sent to the mess. Jammy and I didn't really know Dolly. We knew he wasn't long out of training, been a bit of a character in C Company and now was in Mortar Troop, where not many young marines had an easy life. It would be fair to say Dolly didn't cover himself in glory on that trip. He was an admin disaster in the mess 4L2. Just walking past his bunk became a scramble, at one point someone even considered spearing a flag atop the huge pile of discarded worn clothing and equipment. But there was something I liked about Dolly. Below that veneer of buffoonery, he held within a steely resolve, an ambition that was often mocked by those who could never equal his drive. Resolve to see him become a success in the

Corps and later in the business sector. Resolve to see him be one of the first in the world to row across the Atlantic Ocean. Resolve to survive being hit by a tanker and a hurricane while in a tiny rowing boat somewhere in the Pacific. Resolve to become one of the UK's most renowned and well-regarded maritime and Arctic explorers of modern times. Dolly was, and still is, a walking 'boy's own' storybook and showed one doesn't have to be born with greatness to become great.

After my flirt with death by seasickness on the Bay of Biscay, I'm happy to admit I was dreading sailing across 'The Pond'. Yet the crossing was blissfully calm. The Atlantic became a sea of glass, the only ripples caused by the wake of this grey war canoe and the pods of inquisitive dolphins that escorted us. As someone who yearned for travel just passing by the craggy silhouette of the Azores lifted my spirits. The crossing of eleven days was akin to an Atlantic cruise. Our evening entertainment was limited to drinking CSB for sundowners, playing mess games, and listening to the ship's radio but sharing infinite stories with blue uniformed raconteurs who could make the most dour of spinsters wet her knickers in laughter. Our mornings were spent working up a sweat on the flight deck with more physical exercise, maybe a lecture or two, our afternoons spent working on our tans then getting rather annoyed as our rays were rudely interrupted by the matelots running around in hooded fearnought suits trying to extinguish the frequent fires that blighted this old girl like a dose of the clap.

Like sailors of yore, the ship's company and em-barked force all looked eagerly as the coastline of Florida came into view. Not that we could stand on the passageways gawping for too long. As the crew of a Royal Navy ship, getting into our best bib and tucker to enter Fort Lauderdale Harbour, smartly lined along its superstructure, would be the only way to pull alongside.

Setting foot on the USA for the first time, I marked this personally momentous occasion by going for a run. We'd kept ourselves in shape on board by copious amounts of circuit training but running around the ship made one feel like a one legged duck swimming in circles.

As a reward for our commitment to fitness we then ventured to the legendary 'Candy Store', a bar on the Fort Lauderdale party strip that no longer exists due to its apparent encouragement of impropriety.

In the good old days, lechery was alive and well, so after rebounding from the beach to the bar, the 'Candy Store' seemed rather a welcoming place for it; apart from the fact I was 19, therefore had to wear a wristband to show I was a minor. Feeling like a leper on soda, I rather enjoyed the fact that everyone else could make a fool of themselves and I could look like a well respected pillar of the community, unless of course, I walked into a shop without my shirt on.

We are so culturally different to the USA it is surprising we speak a common language. In the UK, a male can walk around without a top on with no fear of consequence other than ridicule. Indeed it is surprising when you see a t-shirt worn when the temperature reaches 20°C. So with my cultural knowledge as deep as my soprano voice, I was surprised to be rebuked everywhere as I walked innocently around town with my inverted nipples on show. I couldn't get my head around not being able to walk across the road when I wanted and I certainly couldn't understand why a toilet couldn't simply be called a toilet.

Although brought up on a diet of *'Hart to Hart'* and *'Dallas'* in my youth, I was loath to use local vernacular and ask the barman at the 'Candy Store' for the bathroom, which has to be the most ridiculous name for a public toilet ever known to humanity. Is the word 'toilet' so vulgar it has to be masked by such a tenuous charade of a noun? How many times, no matter how upmarket the venue, have you ever gone to the public toilets and found a bath in there? I certainly haven't. Even if there was, I would hardly imagine it being a relaxing experience bathing in a questionably clean bath while listening to and smelling a stranger's piss and shit through a Formica partition, hardly what the makers of *Radox* would portray in an advert. As for calling them a restroom, if I was, for example, in a shopping centre tired of following the wife around the innumerable aisles of women's clothing stores and thought, 'Ooh, I'm a trifle tired I need a rest', I'd seek a nearby bench or even a coffee shop to relax and watch the world go by - young mothers dragging their snotty nosed kid along while continuously knocking them with the numerous plastic shopping bags hanging off the pushchair; or an old couple walking so slow that the group of youths behind pretend to punch them in the back of the head. The last place I would go to rest is a public loo. For one, I would either need to sit on a shit stained toilet that I have to wipe dry due to the moron before me pissing with the seat down; or most likely, lean against the bank of wet sinks covered in foamy soap scum and a collection of questionable skin. Loitering inside a public convenience would almost certainly entail being questioned by security about my intentions, and I really wouldn't blame them for thinking me weird if I answered, 'I am having a rest.' So I remained staunchly limey (apart from wearing my T-shirt) and even if it caused a little confusion asked for the directions to the male toilets, which invariably meant passing the queue for the female loos. Why, despite being in this day of equality are the male public loos always further down the corridor than the females? Is it to give them more gossip time over the porcelain? When we get to middle age those few extra yards can be the difference between walking back out with dry strides or with splashes of water from the sink purposely thrown over our groin area to camouflage the wet patch of dribbled piss we have failed to hold.

The 'Candy Store' barman laughed at my cultural faux pas, and tried to pass on a few local tips to ensure it wouldn't happen again. Stuff that, the next time I wanted a shit, I'd do it in someone's bath.

*'Travel is the frivolous part of serious lives,
and the serious part of frivolous ones.'*

- Anne Sophie Swetchine, Russian mystic

OUR MISSION (The Americans love that word) while in Florida was to undertake our yearly weapon test (APWT), complete our yearly combat fitness test (CFT) and to undergo the US Special Forces basic survival course (SUCKANEGG).

Camp Blanding is a military base so big it's worth taking a map and compass when venturing out to brush your teeth. Set in the pine forests on the outskirts of Jacksonville in northern Florida, we enjoyed our stay, and to a degree were left alone by the US guys, but not the resident mosquitoes. Some of the US Special Forces asked if they could join and complete our CFT. Of course it would be no problem, they were our friends and it was the least we could do to show our gratitude in recognition of their hospitality.

The CFT is pretty straightforward - eight miles of speedmarching in a group, carrying around 25kg of weight, plus a weapon on our backs, running and walking the course. It had to be completed in 1hr 50mins in accordance with British Army Fitness standards. Obviously, we took no notice of what 'Perce' wanted, so would usually do it in around 90mins.

The US SF lads set off with the first group - a troop from Charlie Company. As Support Troop we brought up the rear and set off around 10mins after the first group. Four miles in, we started seeing sweat-strewn stragglers from the front, not one wore a British uniform. At the finish line, only one American SF guy had completed the course, trailing just behind us. Not that they should be ashamed of this failure, their fitness regime trains them for a different aspect of warfare, and speedmarching is pretty much an alien concept. But it still meant they were embarrassed and, to us, liabilities.

We thought we might be a little more impressed by these supposed Special Forces guys on our basic survival course. Even though we'd undergone pretty extensive survival training, we were hungry to learn more from a US perspective. The instructors must have thought themselves as pretty darn good, as us students cooed and wowed in audible fascination when shown Special Forces bivouacs, and windproof matches, trying to impress us with equipment and tactics we'd learnt within the first few weeks of basic training; our perceived adulation that was indeed gross sarcasm. What was supposedly a joint venture turned into joint embarrassment, the US SF guys even apologising at the end when we explained our backgrounds.

A week at Camp Blanding did cut down our R'N'R time, so upon being thinned out, we sped off far quicker than a US Special Forces soldier doing a CFT.

R'N'R from my viewpoint was rather a damp squib. The wide variety of ages in AE Troop meant we didn't have the same outlooks or goals. I wanted to make the most of our short burst of freedom, to do as much as we could in a short a time as possible. So I was particularly frustrated to spend our first afternoon enjoying a Laundromat in Jacksonville, then sleeping in a town where we'd spent the previous week. We did eventually set off the next day, and hopped down the coast to Daytona where we spent the night arguing in a nightclub about what we should do for the rest of R'N'R. Finally ending up in Orlando, we saw Disney World in the morning and Sea World in the afternoon. As those who have been to these resorts can vouch, you can spend a whole day just trying to find your car in those gargantuan car parks.

Despite the plethora of activities on offer, we seemingly had decided to spend most of our time driving leisurely on highways, then rushing around like imbeciles to see things of interest. We did at least spend a night at Church Street Station entertainment area in Orlando, where I walked around with a luminous pink band around my wrist indicating to everyone that I shouldn't really be there getting ignored by every woman despite speaking in an overly-loud voice to hopefully ensnare a British accent lover.

The night ended by Dinger asking our taxi driver if there was an after hours drinking den we could frequent.

'No way Sir, you won't get alcohol now at this time.'

'What about crack?' I joked.

'You want crack?'

'Yeah, I love crack after a beer, me,' I continued.

'Sure. OK, I'll take you to a place I know.'

It was at this point I explained that I was joking about requiring my fill of narcotics, yet this dichotomy did seem a little askew to us Limeys but accepted this as the face of modern USA.

Saying bon voyage to the US, I felt a little under-done by the whole occasion. I'd been so looking forward to it, but with R'N'R being a personal disappointment, I vowed to ensure I'd choose my own company and forge ahead with pursuing what pleasures I could.

Our trip saw us next at the Island of Vieques. A small island belonging to Puerto Rico, it boasts some of the best virgin beaches in the whole of the Caribbean. Locals were split as to the level of hospitality they afforded to the US military who kept a large part of the island as a training area.

Aquamarine seas lapped against the white sand beaches that we used for amphibious landings. This is what the recruitment brochures had in mind. Leaping out into the warm waters here was far more pleasant than the cold waters of the English Channel, with far less turds to wade through. As AEs, we used the demolition ranges to blow up any random object, first having to clear them as the previous occupants had left the area crawling with live ordnance, not something that enamoured us with confidence as we tiptoed around the range trying to avoid a careless loss of limb.

Aruba was our next stop. A little known island belonging to the Dutch Antilles, our destination was the Lesser Antilles archipelago, made up by the ABC islands - Aruba, Bonaire and Curacao. This was the real Caribbean of my dreams. Calypso beats omnipresent, dark skinned exotica paraded the palm tree lined roads, stirring the excitement, naivety and purity of a developing island nation. Tourism was the new boom here and we arrived before the uniformity of branded hotels flooded the island.

Oranjestad hosted our first short stop. If we weren't on the demolitions range we'd be in town, soaking up the Dutch Caribbean flavours or basking on the world-renowned sugared beaches. I started to spend a degree of time with the matelots on my mess deck - a cardinal sin to some bootnecks, yet I really liked the stokers and was happy in their company. As it'd been ingrained into them since they first stepped into Royal Naval training at HMS Raleigh, the matelots loved their rum and what better place than the Caribbean to divulge their interests. It also helped that they were in far worse physical shape than me, so in comparison I looked an Adonis - a skinny Adonis it has to be said.

I continued to share the majority of enjoyment ashore with Dinger and Jammy, and it was with Jammy I spent a warm afternoon on Eagle Beach wedged between the crystalline sea and one of the main hotels. As one does when in such a circumstance, we both felt a little thirsty from sunbathing so decided to partake in a small beverage at the hotel pool bar. Who should be there? Our very own Colours 'J', who was certainly upholding his name of Colours 'Drunk', by quaffing down a bottle of champagne. He was celebrating a big win at the hotel's casino and so shared a glass with us.

'Champagne doesn't go far between three, does it?' he said. 'Barman, let's have another.'

Jammy and I smiled as the barman popped another cork, the swirling vapour escaping from the neck like an opulent genie.

'How much is that again?' asked Colours 'J', a little too late.

'$105, Sir.'

I thought I heard incorrectly. Upon 'J' going for a piss, I looked on the bar bill that the barman put in the small receptacle. I hadn't. *$105 for a bottle of bastard booze.* It was the equivalent of four days wages for me. I'd make sure I enjoyed the experience.

Here I was, 19 years old, sitting at the pool bar of a 5 star resort in the Caribbean slurping on $105-a-bottle champagne. Those days of 5am polishing ablutions floor with boot polish and dragging my damaged body through the foetid mud of the endurance course tunnels all seemed a distant memory. We were continually told in training, that our endeavours would one day be rewarded. As I sat listening to steel band fighting for audibility with the splashing of beautiful bronzed females in the pool, I figured this was that day.

Colours 'J' returned and bought another bottle. After we scuppered that one, he bought another. We'd now downed four bottles, and as the bar bill piled up in front of us, the ship's captain rocked up with his fellow high ranking officers. 'J', now Colours 'Obliterated' decided to buy the skipper and his cronies a bottle to thank them for their hospitality on board. The captain was more than delighted to accept and thanked us for being a great embarked force. I felt a little left out at all the backslapping, so returned to watching the game of pool volleyball played by two teams of skimpy bikini-wearing goddesses.

My mind was so focused on watching bouncing boobs that when I did return to 'planet normal' 'J' wasn't there. Jammy explained that he'd gone for another piss, which under the circumstances wasn't surprising. He must have drunk far more than we imagined because it seemed his visit to the toilet was a long one. Too long for comfort. We waited and waited, then waited a bit more. It was Jammy who first approached the subject.

'Do you think he's done a runner?'

I certainly hoped not. Sat next to us was a bar bill that we worked out to be $525. I went to check in the toilets, as I needed a piss myself. He wasn't there. It was confirmed. He'd left Jammy and me with a bar bill far in excess of what we could afford. Jammy had $20. In my shorts pocket I had 15 wet dollars, some damp fluff and a stowaway seashell. Not quite enough. The only option open was to also leave the scene, so we individually slunk away from the bar, each pretending to go to the loo.

What subsequently transpired I can only imagine, but it seemed to go like this: The barman saw the three now-vacant chairs with a stack of unpaid bar bills in front of him. I'm sure he must have cursed himself for not getting the bill paid in stages and would've dropped a log when he totted up the unpaid amount. His life as a barman would be coming to an abrupt end. However, there was a way to redeem himself. He looked over to the group of gentlemen, sat around a table, who had also been drinking the champagne. They would have to pay. So, imagine the surprised look of the HMS Intrepid Captain when given the bar bill for $525.

He may have been a switched on cookie when it came to explosive engineering, and a fine leader of men who never asked of them what he could not do himself (unless it was to pay for champagne); but 'J' was the world's shittest criminal. The captain returned to the ship incensed, possibly due to him being escorted by the police when refusing to pay the bill. 'J' was crudely awoken from his champagne-induced coma by the Joss - the head of the Naval Police on board - and ordered to pay the outstanding amount. Lucky not to be thrown in jail by the police and even more lucky not to be charged by the captain, 'J's punishment was to be made 'Duty Booty' while the ship was alongside for the remainder of the trip.

Nights out in Aruba were pleasant to say the least. A haven for rich American tourists meant we could mingle freely and not have to try too hard to meet easily impressed females who, in holiday mode, seemed to leave behind their normal standards of decency. The enjoyment we sought was countered by the tragedy that struck the ship yet again. One young sailor on his day off frolicked in the shallow waters off Aruba. Diving into the sea he hit his head on the sea bed, breaking his neck. Paralysed from the neck down, he floated head down in the water until help arrived. He was flown back to the UK soon after, an adventurous life shockingly cut short by something so inane as shallow water.

**

Willemstad, the capital of Curacao, is reminiscent of a sun-drenched Amsterdam, where Dutch-influenced pastel buildings lined our arrival in a multi-coloured welcoming parade.

The fighting company lads went off to do their soldiering and we again stayed on a demolitions range blowing up any piece of metal that was lying about. Even with Colours 'J' banned from going ashore, he felt it important to still finish at lunchtime and give us ample opportunity to sample the touristic delights of such a beautiful island.

Possibly the most attractive destination for the sailors and marines was 'Campo Alegre.' Literally translated as 'Camp Happiness,' a large open aired brothel, it traded quite legally inside its floral boundary.

Servicemen and prostitution go together like spotted dick and custard. Having seen the fallout of many post coital business transactions this is a rather apt analogy. When away for months at a time with only the pages of a well used magazine for love, when let loose into a general public plentiful of women who are eager to please, morality is placed in a locked box and replaced by lupanarian intent. To many males, there was little regard placed for the welfare of why and how these women had ended up in such a profession. To many servicemen, trapped in a male dominated bubble, the suffragette movement was something to do with abnormalities of an aircraft engine and their focus when abroad was only to get their nob dirty. Brothel takings always go up when a ship comes alongside or a military unit arrives and their intelligence network is so accurate, the working girls unerringly know when the first lads will arrive. Terrorists could save so much time in reconnaissance and information gathering by just seeking out the local brothel owners and they could mount a perfectly timed attack.

Men, with previously nothing to spend their earnings on are now awash with spare cash, and there is only so many fake Rolexes you can try on, so beer and women seem to be the next items on their 'to do' lists. In many countries you have the convenience of establishments selling petrol with a side shelf of overpriced groceries. In Curacao, like various other places in the developing world, you have the convenience of beer sold next to rows of underpriced women. Campo Alegre certainly satiated the needs of those with parched throats and overloaded testicles.

I went along with Jammy, in the name of voyeurism, to see a prostitute in the flesh. I'd only ever seen them on the TV and that was only when one had been murdered by Peter Sutcliffe.

Campo Alegre was similar to a military camp with small walkways of horse box like rooms, each one presented with a beguiling lady, lurid from lumpy foundation and runny mascara, dressed in various forms of little, cooing to the visitors with offers of products and services that couldn't be found at your local Tescos.

'Hello sailor boy,' she purred in my general direction. She had the look of an Amazonian Goddess, dressed as if she'd just got out of bed. She probably had.

I looked around. There were no sailors, just two shit bootnecks - us.

'Yes you.' She extended her chocolate coloured finger in 'come hither' gesture. Thankfully it was the sultry exotica of her skin tone, not where it'd been tickling the prostate of a drunken sailor.

'Hello,' I stammered, not knowing how to converse with a woman of such occupation.

'You like what you see?' She (we hadn't yet been formally introduced) slowly span around.

'Yes, it's very nice,' I lied, as if answering a car salesman.

Through the lilac sheen of her polyester negligee which, to be honest, wasn't appropriate in such heat, I could see the ravines of mocha stretch marks running through her shimmering skin. She did have a nice arse though, it had to be said, which was a nice contrast to her varicose veins, which must have been a genetic disorder as I imagine she spent most of her time on her back.

'You like to see more?' Her lilted voice became huskier as she slowly peeled down her yellow bra to reveal the top of her dark brown areola. With the breaking strain of a *Kit Kat*, I was hooked.

'Err...'

'I thought we were here to just look,' said Jammy from outside my field of zoned vision.

'Yeah, I'm only looking,' I said truthfully. My heart started to beat in nervous anticipation.

Within a blink of an eye, I was at the doorstep. She held my hand. I instantly regretted wearing tight shorts - my excitement was physically palpable.

'Come on baby, step inside,' she purred in a very sexy voice. 'You have American dollars?' she stated in a very bureau de change voice.

Her room was fragrant - fragrant of cigarette smoke, cheap perfume and spunk. It was like being back in a Northern Ireland sangar. I peered in, and my head instantly sweated from the musky heat blown down by a pathetic ceiling fan. Inside, the sauna-like room were the tools of the trade of a working girl: a stained bed, lamp and a bedside table full of lubricants, condoms and toys. She was organised - she even had spare batteries.

She bundled me into the room as if I was a dignitary escaping an assassination attempt.

'I just want to look,' I said, realising my surroundings were about as sexy as an abattoir.

She paused eyeing me up. I was young so probably didn't have the money she wanted for a full servicing. 'OK $10 to look,' she said hands on hips looking like Wonder Woman gone rogue.

'Timey, we going or what?' said Jammy from outside.

'I have to go,' I said pathetically.

'$10 to look,' she said. Her spellbinding smile had turned to a spell conjuring snarl.

'Fuck off, I'm not paying you $10 to look at your room,' I said. It was a fair one - the Natural History Museum in London is far better value.

She then rattled off something in, I assume, Papiamento dialect. It was responded by a voice that would have been well apportioned to a 6'5" monster.

It was. He filled the doorway.

I was Royal Marines boxing champion in my day; I could have him. His baseball bat he produced from his back told me I couldn't.

'$10?' I said huffily. I pulled out the most crumpled note I could find, just to exaggerate my impotent belligerence, and threw it onto the bed.

'Go on, run to your boyfriend,' she said as the human door opened.

With a face like thunder, I looked at Jammy. He started laughing. I followed suit, not stopping until we left. I was the world's shittest curb crawler.

With emptied scrotums, the mood of the ship lifted as we returned to Aruba for a quick exercise before heading off to Barbados. Here Charlie Company group would split. Our group would spend two weeks in Barbados but have to work the duration. The other half would go to Grenada for a week's exercise then return for a week off in Barbados. It sounded like a fair plan and so with Support Troop now living in the Barbadian Defence Force (BDF) Barracks near the airport, we found with no one really in charge we succeeded in doing as little as possible for the first week. We were in Barbados for God's sake. Who in their right mind would want to work themselves to the bone? We did some token weapon training with the BDF and each morning we'd run off our hangovers down to the beach, a haze of rum fuelled vapour hovering above us, beer farts emanating below, before jumping in the water to have a bit of a sea swim then return to get our beach rigs sorted. This would seem an exhaustive and laborious task as it somehow took us all morning. We did manage to finish coincidentally in time for the BDF's first shuttle run of transport to take guys to Rockley Beach or further into the capital Bridgetown.

Prepared with our night clothes rolled neatly into our daysacks (a military man should always be prepared) we would sunbathe, surf Rockley's famous waves or meander around the hotels. Should we wish, we could have an afternoon taking in the sights and sounds of vivacious Bridgetown, or take a land rover up into the surrounding hills and explore the plantations. In short, the first week was a holiday - a swan, and not one fuck was given.

As the street lighting took seniority over the sun, we used stinking public toilets to change into our night clothes and frequent the many bars and hotels that were happy to take our not-so-hard earned money.

A night trip on the famous *'Jolly Roger'* would be the finale to our first week. The rest of the company would be joining us the next day, meaning an unwelcome week of pseudo discipline and, God forbid, work.

The *'Jolly Roger'* was a party boat branded itself as a remake of a buccaneer ship recreating the exploits of pirates enacted through sailing into the blackness of the Caribbean Sea while the onboard passengers would dance and drink to excess. Possibly to reflect the bawdiness of those days on the high seas, they allowed 20 or so toga-clad marines on board. The crew would be correct in thinking we would be as true to historical lore as possible - save our dress from antiquity. Adopting the psyche of Black beard's most twisted crew, we climbed the rigging, jumped overboard the moving ship, and chased screaming girls around the upper decks. Activities, it seemed, that were all prohibited despite our protestations that our antics were purely pirate-like, even promising to stop short of beheading landlubbers.

The rest of Charlie Company group returned. They were fresh from a bi-nation exercise in Grenada in which the US Navy SEALs had thought our lads were SBS Special Forces and not actually baggy arsed bootnecks through managing to scupper every SEAL operation on the island. After a week in the jungle, they too were looking forward to some ship borne merriment in celebration of their victory over our noisy neighbours. However, our naughty pirating had come at a cost. Everyone from HMS Intrepid was banned from the *'Jolly Rodger'*. Banned from a pirate ship? How badass were we? The ban wasn't particularly well received by the guys. The ship's radio DJ, a young matelot who, for the best part did a sterling job of entertaining while at sea, publicly slated the lads over the tinny airwaves. Not the best career move for someone so young, he was later found tied up in a seaman's kit bag, swinging in time with the ship's roll, from the crane jib that hung 20 feet above the flight deck.

As feared, the second week was more regimented. We actually had to work, finishing at the ungodly hour of 1400hrs. We did, however, still manage to find plenty of time for recreation, although we did think it a liberty having to do a night exercise.

In true commando style we were delivered by rigid inflatable boats (RIBs) to a disused power station, where we would steal in, unbeknown to the BDF guard force, then simulate blowing the shit out of it by the clandestine laying of dummy charges. These improvised charges were clusters of thunderflashes connected by Fuse Instantaneous (FI) with a length of safety fuse to initiate. These would make a hell of a bang and would signal the destruction of the power station. The BDF soldiers didn't really pick up on us as we placed the charges. They certainly didn't see us leave. They certainly heard the six loud bangs as the charges detonated, and we watched with chuckling glee from our offshore vantage point in the RIBs as they ran around like a mad woman's piss to try and catch any stragglers. The problem was, we only heard six bangs. There should have been eight, which meant we had two blinds to contend with. We couldn't leave them as they were live ordnance and quite dangerous, if one of those went off in your hand, our pirate impressions would have become far more realistic. So as soon as the exercise was deemed over, we waited the allotted time before retrieving the two blinds to dispose of as necessary. It would have been simple to throw them overboard as we returned to the landing craft but Bully would not have any of it. He would dispose of them in his own inimitable style.

Bully was one of those characters well known in the Corps. Unfortunately it was usually for acts of crass stupidity - he was an '*idiot savant*', without the savant. Undoubtedly a child of Imber, he was now an old warhorse having served nearly 22 years but had only reached the dizzy heights of corporal. Not that this is necessarily a bad thing for someone who wanted to stay as a foot soldier, but Bully wanted promotion and so only attaining this rank was a reflection of his competence. He was the 'Reg Prentice' of the Royal Marines and Kenny Everett's fictional character was a perfect send up of Bully, a continual sausage of disaster who could turn a simple, safe job into a near death experience. He was the butt of many jokes and Dinger had managed to get him on the old classic trick of changing his watch to 8am while asleep then waking him telling him he was late for parade. In the bowels of a ship with no sunlight to guide the body clock Bully, panic stricken, changed quickly into his rig rushed up top onto an empty flight deck only to realise it was actually 1am.

Furthermore, in response to him falling asleep by the roadside while in Sardinia, in anticipation of more daytime slumber on the next exercise in Aruba, we emptied his packed bergan, replacing the contents with pillows. As we were good eggs we didn't want to see him get fat by just sleeping, so Dinger threw in a couple of weights to help him keep in shape that, by coincidence, was similar in weight to his original bergan. A week in the field, even in the mildness of the

Caribbean, is a long time if you only have useless pillows and two 15kg weights as your kit. It would have been easy for Bully to ingest all these claims of idiocy when considering his actions, yet he had the skin of a rhino, so continued unabated.

On getting rid of the unexploded ordnance from the power station, he would certainly do it his way. Mark 8 thunderflashes hold a fair amount of black powder - a low grade charge. Think of them as an overgrown 'banger' firework - the staple of many a teenage boy between mid October and late November to the dismay of anyone within a mile radius of their tomfoolery. On its own, a Mark 8 thunderflash makes an incredible racket and the explosion can burn those nearby. When many are taped together with FI connecting them all they are quite a powerful pyrotechnic (and also an illegal one). We'd taped four of them together for each dummy charge. With two charges now unexploded we had eight thunderflashes to get rid of. With the normal strikers removed and replaced with FI, Bully couldn't ignite them normally. I, as the least experienced, retreated from the debate allowing the more senior lads to discuss the disposal methods available. I can't envisage anyone agreeing with Bully on his decision. As I passed him again, he was emptying the black powder from the thunderflashes into a small conical pile.

'What you doing Bully? I asked.

'Making a genie,' he replied.

'How old are you, nine?' I said.

My stepfather once made a genie out of an emptied banger at that age and it nearly took his fingers off, so I shook my head thinking he was joking. Not even Bully could be so stupid. I returned to my first floor balcony and started polishing my dusty boots. In an instant, the air below me took on a bright orangey red glow as the fizz of lighted black powder broke through the air. I looked over the balcony. Below was Bully walking from a cloud of smoke like a cartoon character, his hair smouldered above a face red from where it had lost skin, his hands singed to a light frazzle. Burnt hair mixed with the sulphurous egg smell of ignited black powder hit the nostrils. It was none too pleasant, but hilarious with absolutely no sympathy for the idiot.

According to onlookers, who I cannot doubt due to them being able to concoct such a story, Bully had first made his gargantuan pile then created a powder trail. If you were to be so foolish as to attempt such a heedless pantomime then the only option was a luxuri-ant trail, duly long enough to afford time to retreat safely from this newly created Frankensteinian incendi-ary bomb. Not Bully, his trail was so short that he was virtually stood over the powder cone that now reached calf height. To add an extra cowboy element, he'd apparently lit the trail with his lighted cigarette, I imagine taking an overly long drag on the stub before smiling smugly to his audience before bending over to touch thunderflash armageddon.

Civilian thunderflashes do have a warning directing owners to keep away from 'direct sunlight, dampness and children'. They forgot to add 'Bully'.

Launched from a Landing Craft Vehicle and Per-sonnel (LCVP), our next amphibious attack on the beach adjacent to the Island's most expensive hotel, raised some very posh eyebrows. Instead of wading ashore with weapons and munitions, we were armed with cool boxes, ice blocks, charcoal and half-cut oil drums.

This was our end of tour barbeque where we could relax and regale on the tour's successes and failures. Jet bikes were hired and I went out speeding about with Jas and Bobby, two lads who weren't in my troop but with whom I'd built up close relationships. (Jas and I once jumped on a train from Taunton one Wednesday lunchtime with a crate of beer for company with our destination undefined until ticket collector would ask why we hadn't any tickets. On reaching his home town of Portsmouth, we went into the Park Tavern pub for a couple of drinks before heading back to Taunton. We did this for no other reason than it was 'something to do in the afternoon'). As a trio, we zoomed about with gay abandon trying to get as close to each other as possible without causing an expensive crash. One holidaymaker had fallen from his jet bike and floundered in the water. His jet bike had floated away.

'Hey guys, could you help me out the water and help me get my bike?' he said with a Southern US drawl.

'Yeah sure no probs,' answered Jas. He cast a mischievous glint to Bobby who passed it onto me.

'Here mate,' said Jas, slowly approaching him, hand outstretched. He then turned the handlebars, and sped off full throttle covering the floating holidaymaker in the white water bow wave.

It was even funnier when all three of us did it. We were like the 'Red Arrows', but on jet bikes and not as good. As Bobby and Jas sped along to cross each other, they seemed to miscalculate their angles. Bobby turned. Jas didn't and sped straight into the side of Bobby who flew high into the air. Bobby came up from the white water, laughing. His hilarity halted once he realised the sea of red around him. I quickly dragged him onto my jet bike and saw his thigh carved open like a joint of ham. His leg dangled at an unnatural angle with lots of red stuff coming out of it in a particularly worrying way.

If he'd indeed broken his femur, he could bleed out, so I sped back to shore as quick as my jet bike would carry us. As always, the medic accompanied us. As always, his relaxation was interrupted by tending to injuries sustained by idiotic bootnecks partaking in moronic activities. Bobby was flown back to the UK soon after; his broken leg containing more stitches than the Bayeux tapestry.

Barbados was the last stop on our deployment. I'd been to places I'd never before heard of, I'd cruised around some of the world's most beautiful islands and laid on their pristine beaches. I'd eaten lobster, drank champagne meeting fantastic locals and over-friendly female tourists along the way. And I'd been paid for every minute of it, although my bank balance would suggest I'd spent more than I'd earned. I'd struggle to throw any cash around in Turkey where I was shortly to go on holiday.

The eleven day crossing back to Blighty passed far too quickly. There is little that confirms more the wonder of nature than the golden sun slowly casting a lengthening iridescent reflection into a silver sheened ocean where no land can be seen. With a warm can of beer in hand, and the ship's engine murmuring in the background, I could have stayed on board forever.

'We are born. We die. Somewhere in between we live.
And how we live is up to us. That's it.'

~ *'Tell Me When I'm Dead', Steven Ramirez*

AFTER BEING ENVELOPED in the vivacious colours of the Caribbean, the dullness of the grey and greens of home weighed heavily, as did my mind. I put my listlessness down to post deployment blues and to escape attention from those who'd ask why I was so low, I returned home to spend a week in bed, not having the energy to do anything or go anywhere. Motivation is the engine to the soul, yet it had disappeared along with the ignition spark that so often woke me up in the morning. Mum, none the wiser, due to our detached relationship and her workaholic nature, accepted my excuse that I was so tired from the deployment I needed a week in bed to decompress. And maybe I did. But this wasn't decompression. My head had closed like a bankrupt shop. Black clouds surrounded my thought and sleep became a battle of dark rumination. Self-starvation became my silent cry, with a bad stomach from a bug 'I must have picked up in the Caribbean' my cover story for not eating. With nothing other than interrupted sleep as my medication, self-loathing pushed me further away from confusion, directing me to a tunnel where I knew for sure I wasn't fit to be a human. What was the point of it all? Why would anyone care for me when I couldn't care for myself? Beyond my bed, only fear stood until it dispersed. Time was not measured by the clock but by the relief when this dark ghost slipped from the hole in which it had crept. The euphoria of waking up with a smile on my face made me feel a million dollars.

For now, the where's, why's and how's of this possession could wait. I had a holiday to take.

***

Returning from leave, we returned to blackshod training. Slate grey streets became littered with burnished oak leaves and the cooling winds mirrored my fear that an autumn of discontent would be upon me. Conventional warfare was becoming quite bland. Training to keep our skills was irrefutably important; however, I did find its repetitive nature less than stimulating. Even in AE Troop, where we spent stimulating days setting up pyrotechnics for battle simulation, I felt there was more to soldiering. We still supported the conventionality of the fighting companies, who patrolled, returned to their harbor position, then moved to do it all again, only from a different location; the only break in monotony being someone walking into an electric fence. I often couldn't wait for such exercises to finish and was no different when we finished on this particular Friday morning. By the time we'd returned, showered and returned back to the AE Troop store, the harrowing news had filtered in.

The IRA had detonated a bomb at the Royal Marines School of Music in Deal, Kent. Reports were unconfirmed but a number of bandsmen had been killed in the explosion. The news sent shock waves around the camp.

Modern day Royal Marines Commandos have a very strange relationship with the Royal Marines Band Service. Commandos rarely came into contact with the 'bandies'; indeed up until the Deal bombing my last sight of a bandy had been at my passing out parade over two years before.

There is a certain caste system within the Corps where bandies aren't classed as proper bootnecks. They belonged to the Royal Marines for sure, but they weren't commando trained and their military training was hardly robust, so were seen as lesser people by many. But as is often the case, it is only when something as tragic as the Deal bombing happens, do the ties of brotherhood strengthen.

The anger was directed at PIRA not only for them attacking the Royal Marines, but the cowardice of attacking bandsmen. These guys weren't combatants. In times of war they may get a walk on part as a stretcher-bearer, but their job was to give pleasure through music. To attack them was like a 6ft rugby player attacking a kid wearing calipers.

Back on Norton Manor Camp, as I was getting my laundry finished, the wandering guard patrol caught me folding my field underpants. I was to report back at the Globe Theatre in figures ten.

I sat in the theatre with about forty others. Most of us were inliers with nowhere to go - the 40 Commando 'orphans'. We were to be immediately deployed to assist with the relief effort in Deal. Quickly grabbing the necessary clothing and equipment we boarded the bus to Kent.

It was dark upon our arrival, but the spotlights shone on fluttering yellow cordon tape, reflecting light onto the rubble of cindered debris where once stood the recreation block behind the entrance to the North Barracks on Canada Road.

Although what I saw was horrendous, it didn't re-ally affect me. Was it the training I'd received? The fact I'd already completed a tour of Northern Ireland? Was it I'd worked around explosives for the past year? Was it I'd become hardened to the effects of inhumanity, through all I'd endured both through childhood and my time in the Corps? Or was it I just had to get on with the job in hand? Deep self-analysis at such a young age, as I'd already found with my dark secret, may become burdensome rather than a solution, so accepted the scenes I now witnessed. Thinking about things could wait until I started sucking on my salmon paste sandwiches in an old people's home.

We were briefed in the galley next to the fallen building. Eleven Bandsmen had been killed, a further twenty one in hospital with serious injuries. Our job was to supplement the security footprint of the camp, conducting patrols in and around the camp - talk about shutting the door after the horse had bolted. Moreover, we were there as a confidence booster for both the bandsmen and the local populace. Briefing complete, we were split into watches and immediately set about securing the area.

Amazingly, we patrolled the streets around the camp fully armed. It was like being back in the Province. Under normal circumstances, to patrol the streets with weapons on the mainland was prohibited, the only weapon the general public would ever see would be a barracks main gate sentry, but here we were outside the camp perimeter patrolling as if in Belfast.

The world's media was upon us. The Duke of Edinburgh was to make a visit; therefore, we needed to be further on our guard. As well as our overt patrolling, covert armed operators roamed the streets to relay any suspicious activity and the more discrete security services were present to ensure his visit went without a hitch. As I'd been on night duty, I was in bed when 'Phil the Greek' came to pass his condolences and show his support.

As Captain General of the Royal Marines, Prince Philip was also affected. With his often-flawed honesty, he was looked upon with a certain amount of affection within the Corps, and we appreciated him visiting even though the underside of my duvet was more important.

What did strike me was the outpouring of grief from the locals. Rows upon rows of flowers lined the South Barracks gate further down Canada Road, and many came to give personal thanks to us for just being there. Many a local came to stand silently at the improvised wall of remembrance, even visiting children wept openly. For the press this was a good story. Unfortunately, the union of ethics and journalism is sometimes dissolved, one of the bootnecks catching a journalist asking for a child to cry in front of the rows of flowers for a good photograph. The bootneck, with little tact but a lot of swearing, instructed the journo, in no uncertain terms, to disappear or he'd ensure there would indeed be someone crying.

With the Prime Minister visiting the next day, more preparations were at hand. Many hours had been worked over the previous two days, so with a few hours off it was only reasonable for us to go into town for a relaxing beer. It was impossible to have just one. Wherever we ventured, the locals, knowing of our presence, bought us all a drink of gratitude. Between us, I would have been surprised if we spent a quid. The young ladies of Deal were also welcoming, and with my defences down, once more I found myself waking at 4am, freezing after falling into a post coital sleep on a park bench. Totally lost, and with a sentry duty in two hours time, I left in search of a taxi. At 4am on a Monday morning in Deal, a taxi is a rare commodity, especially in the back roads of suburbia. I decided to head east to hit the coast then find my bearings, but it was still pitch black and there were very few lichen-covered trees to find my cardinal points. Astral navigation was out of the question on a cloudy night, so I jogged down suburban roads hoping to find a recognisable reference point in a town I'd seen very little of. Luckily, I bumped into a milkman in his float doing the early rounds, so with a free bottle of milk and directions back to camp over two miles away, I ran back, heading straight into the shower and back out again like Mr Benn ready to go on the South Barracks main gate.

In truth, I'd have preferred to go to bed, I was knackered and standing on a main gate for two hours is not a great hangover cure, especially with the smell of vagina still hanging around my top lip. However, with Margaret Thatcher arriving it was an even more ridiculous situation to be in. I just hoped she bypassed me. She didn't. With my position next to the Wall of Remembrance she made a beeline straight for it. There was I, trying to cover my arcs with the world's media in front of me hoping their cameras couldn't pick up on what shit state I was in. I scanned the rooftops and attempted to look professional. Maggie approached. The Prime Minister of Great Britain and Northern Ireland looked at me. 'I'm sorry for your loss,' she said softly with genuine concern.

I'd have rather kept scanning my arcs at such a critical time but when the PM speaks to you I suppose you have to look them in the eye. *Fuck*, would she smell my beer breath? Would she smell the vagina odour emanating from the gutter of my philtrum? I must have smelt like a brewery toilet.

'Thank you Ma'am,' I replied quickly through gritted teeth, not even knowing whether calling her 'ma'am' was correct. I returned to my arcs, casually nodding to some high ranking official following her, attempting to prove that protecting the PM was far more important than military etiquette.

No sooner had the relief of not killing the PM with my breath was over, I was out on the ground again with an old chef called Dicky. He'd completed the first phase of his Royal Marines training here when junior troops spent their first 12 weeks at Deal. Having personally seen the support for marines, bandsmen or otherwise, from his first day as a bootneck, he was overwhelmed when both him and I led the bandsmen from the North Barracks to the practice suites. The band marched in formation playing 'A life on the Ocean Wave' - the official Royal Marines regimental march. As we patrolled ahead of the band marching proudly down the street, people left their house and applauded from their front gate. Old men stood in the street and saluted; young mums pushing prams stopped and encouraged their babies to look at us. My spine tingled with pride as one old man, gnarled from age, yet with a youthful twinkle in his sparkling eyes said, 'Well done Royal' when we walked by.

I looked across to Dicky. He had a tear in his eye. I steeled myself. I'd previously felt no connection with these bandsmen, but their effect on the locals, and the overwhelming support showed me they truly were fine people, and as the defiant drums beat to the sound of powerful brass instruments, I felt honoured to walk ahead of them. The band would not let such a cowardly atrocity stop them doing what they loved: play music. It was all they wanted to do.

***

My new year resolution was to find new challenges, and there is little more challenging than spending six weeks mountain training in Garelochead, Scotland, especially after spending Christmas leave travelling around North Africa.

Mountain training wasn't something that particularly interested me. It was our bread and butter, yet this metaphor couldn't hide that I was, yet again, getting restless and wanted something completely different. Being in AE Troop was too much of a doddle and one can easily get bored of darts and fixing toilets. There had to be more to life. The answer to my prayers came as I read the Daily Routine Orders that were promulgated, as normal, outside the CSM's office. DROs had to be read, it was a chargeable offence not to, but hidden amongst the normal drivel, sometimes gems of information were uncovered and on this particular cold and dreary day it came in the form of a paragraph entitled: 'Volunteers For Belize.'

7 Battery, part of 29 Commando Royal Artillery, were short of men for their upcoming tour to Belize and needed ranks to assist their manning levels. Deployment would be in April but volunteers would be required at the beginning of March for pre-deployment training. It sounded perfect. As I gazed at the fog-shrouded hills amongst the misery of south eastern Scotland, I knew six months of Caribbean jungle would revive my soul and keep my North African tan up to scratch.

Even though 29Cdo belonged to 3 Commando Brigade, I was warned that I was venturing into pongo land. I'd never really met any 29Cdo guys. Even during the brigade exercises we shared we rarely crossed paths, so I was worried by ignorance rather than by design. By stepping into Army territory I didn't really know what to expect. Would I be marched around by some archaic drill instructor barking orders from the guardroom to the jail for some minor and irrelevant offence? Would I have to call a corporal 'Corporal', a sergeant 'Sarge' with my heels together like a nod? Would I have to call scran 'scoff' a wet a 'brew'? Would I have to slam my feet into the ground and splinter my shins whenever I stopped marching like a guardsman? Only time would tell.

I was excited. Toni, my new girlfriend, wasn't. I'd met her prior to Christmas leave. We'd spent a total of five nights together before I headed off to Tunisia. She understood that; it was booked. She even understood going to Scotland immediately after leave; it was planned. But I'd volunteered for this trip, volunteered to be away from her for six months. The early Easter leave I was granted meant the hastily booked romantic skiing trip to Andorra, where I could venture off on my own while she stayed at ski nursery was a holiday of indifference rolled with the air of resignation that rises when a relationship is clearly coming to the end of its natural life.

It was with eagerness that one only gets with travelling to a new destination that I boarded the night train to Arbroath, Scotland. My partner in the sleeper was Pete, someone I recognised from C Company. Hardly saying a word as we sorted ourselves for the night, I thought him a little aloof, especially as we were bootnecks going into army territory. I didn't push things, he may have recently had a relative die, he may be thinking of a loved one he'd left behind, or he could just be an ignorant fucker. Nevertheless, we bedded down for the night, arriving in the early morning, the east coast Scottish air as cold as the preceding scenery bland.

If my previous worries of 7 Battery being a haven for army barmy pongos with a penchant for military discipline and strict regimes, I couldn't have been more wrong than if I'd asked Charles Manson to babysit.

The 7 Battery Sergeant Major (BSM) called us into his office. With his feet casually on his desk, he introduced himself by first name. He was chuffed that we'd volunteered for the tour, and hoped his lads could learn as much about soldiering as we could about gunnery. Not that theirs needed much improvement. From our first lessons with them it was clear that these lads knew soldiering.

Outside of military skills some were wary of us newly arrived bootnecks, some welcoming. As they were permanently co-located with 45 Commando Royal Marines, the relationship between the two was sometimes strained, so I didn't expect the open arm treatment from everyone. I reciprocated the goodwill of those who went out of their way to welcome us and stayed out of the way of those that didn't. I was their guest and didn't want to spoil my invitation from the start. Besides, some of them were big fuckers.

I was roomed with a gunner called Geordie. A really friendly lad but one that clung exceedingly close to me as soon as I arrived. His limpet-like friendship and him having a four-man room to himself sounded alarm bells. To say he was slightly strange is bit of an understatement. He would fart and cup the smell to his nose; an action that I think is referred to as a 'cup cake.' He would blatantly cup cake regularly. He would also masturbate like a grinning zoo chimp and far too often. As his bed head was butted against the back of my locker that separated our bed spaces, the tell tale sound of my lock swinging in unison to his trembling movements meant my Walkman got extra use. I was later told that on one occasion when embarked on ship, an emergency drill was called. His bed had its curtain pulled closed as often they are, but as the lads departed the mess deck nothing moved from Geordie's bed.

Flinging back the curtain to check he was there, a gunner found Geordie with his Walkman on full blast, sprawled naked, wanking wildly to AC/DC.

The four weeks up in Arbroath were enjoyable. I'd bumped into some old oppos and even met Jim Davidson, who was the first person I'd ever witnessed grand slam his bed whilst we'd shared a room together in training. He was also the first man who I'd ever seen ride a motorbike, drunk and naked down four flights of concrete stairs being chased by the 45 Commando Duty Sergeant, at 6pm. Jim would go on to become an RSM.

The beat up in Arbroath complete, we finally deployed to Belize. The lads I had newly met, both army and marine, were quickly becoming trustworthy brothers. I was off to the Caribbean again, this time for a full six months. Excitement didn't cover it.

***

Squeezed against the azure Caribbean Sea by Mexico to the north and Guatemala to the south and west, Belize became another British conquest and was renamed British Honduras until 1973. The country claimed full independence in 1981 bringing with it vulnerability from the relatively strong military might of neighbouring Guatemala, who claimed irredentist rights over the land on which Belize still stands. Despite its sovereignty, the Belizean government accepted the dependency of their former ruler for security, so an agreement was made for the British forces, who had secured a continual presence there since the 1940's, to remain there as a protection force. This was highly agreeable to the British. Not only did it mean a continued presence there (any excuse to still retain one foot on foreign soil), but also a fantastic training ground for its troops and air force to hone jungle skills, quick reaction operations and air mobilisation.

Manned by an infantry regiment, a Royal Artillery battery, a cavalry regiment and a Royal Engineers squadron plus administrative add-ons, the British Army had the largest presence there. A Royal Air Force helicopter squadron and a couple of Harriers made up the resident units. The Royal Navy, other than stopping off to allow matelots to visit brothels, had little of a presence and as the British Forces Belize rotated its infantry units similarly to the rotation in Northern Ireland, Royal Marines commando units, coming under the leadership of the Royal Navy, got little of a look in. Funny how we seemed to be permanently on the list for the wet and cold, dangerous places; yet for a six month tour in the sunny Caribbean, where coconuts falling from palm trees would take the place of coffee jar bombs thrown from blocks of flats, commando units seemed to have a rare presence. 40 Commando had previously been stationed there in 1985 and it would take another eight years for Royal Marines from 45 Commando to experience a full tour. This strange anomaly also occurred for the routine UN tour in sunny Cyprus where only 40 Commando had ever ventured - seemingly getting a UN medal for raving it up in Ayia Napa was the exclusive remit of the British Army.

In truth, while the jungle operations and commitment to Belizean security was important, a six month tour there was seen as a bit of a 'swan.' It's not often you spend six months in a country of such diverse ecology and natural beauty that within a couple of hours of each you can swim, snorkel and dive on the second biggest barrier reef in the world, laze on sugar white sand beaches, trek (Or as we did patrol) through untouched jungle, hear the tubular roars of howler monkeys and even spot rare jaguars, see ancient Mayan structures, and recreate a winter's day on Dartmoor should you want to visit Baldy Beacons.

Disembarking from the RAF VC 10, we bathed our faces into the warm humidity of the Central American sunshine, breathing in the wonderful aroma of a new country - the fillip to even the most beleaguered traveller. We split into our respective groups. Half the Battery was sent south to Rideau Camp. I was in the group sent to Holdfast camp a 60 mile journey from Belize City along the Western Highway.

Along the way stood JB's bar. Famed for its Belizean hospitality towards the British forces, stopping there was a rite of passage. A coat of arms from every touring unit decorated the walls, and a place where probably every serviceman headed west drank their first Belikin beer of the tour.

Belikin Beer is a strange mix. Brewed at the gates of Airport Camp, the British Forces Belize HQ; each bottle tastes slightly different. On a good day you could get on a roll where every bottle tasted pretty good. Unfortunately, and far too frequently, a bottle would be opened that tasted as though it had been strained through a pair of used incontinence knickers. The positive to take from this was if feeling extremely ill or hungover the next day you could easily blame it on a bad Belikin. Over the next six months I would drink too many of those.

At JB's I, like countless others, bought the very essence of what a tour t-shirt should be, funny, yet offensive to some, and stained everything when first put in the washing machine.

It read:

I DON'T SMOKE DOPE CHEW ROPE

DANCE, PRANCE, ROMANCE, FIGHT

FART, FUCK, SHOOT THE SHIT OR DRIVE A
TRUCK

I'VE BEEN IN THE ARMY ROYAL MARINES

ROYAL AIR FORCE AROUND THE WORLD

THREE TIMES, THREE WORLD WARS AND

SEEN GOATS FUCK IN THE MARKETPLACE

BUT I AIN'T NEVER SEEN NO SHIT LIKE THIS

One could class this T-shirt as an antique, taking its place at clothing museums in between royal wedding dresses and Henry VIII's feathered velveteen hat. Although I still can't overlook the treble negative on the bottom line.

Our accommodation at Holdfast Camp was one of the many old Nissan huts surrounded by a moat of deep monsoon drains. My creaking iron bed was conveniently next to the exit into the Battery bar - a grass roofed atap that would become the centre of many a night of alcoholic debauchery and home to some big fuck off tarantulas. Being the first night, it was only sociable to fill, then empty the beer fridge and make the sparse bar a more homely place. The bar soon became crowded, and much mingling and merriment commenced.

Although the first night, one bootneck, called Mac, was already receiving sexual attention. Unfortunately, it was from another colossus of a marine, also from 42 Commando, called Johnny, who decided to put down his own mark and decide to give Mac a blowjob on his flaccid penis. Mac could boast he was the first one to get a blowjob on tour; the caveat being it was committed by a fellow bootneck. At least it wouldn't be classed as cheating by his girlfriend. One can only imagine that when receiving such attention it is always preferable to take your mind off what is happening below. Mac seemed ambivalent as he held a can of beer and talked to another about subjects that were personally distasteful. God forbid Johnny be good at it, I'm sure the last thing Mac wanted was to show physical proof of any enjoyment. Yet this was Johnny and Mac's marker - their way of showing that bootnecks, especially those from Plymouth, would spend the next 6 months pushing boundaries - although personally, such activities were beyond my remit.

Within a couple of days, half the lads were away into the jungle to undertake their two-week jungle training. Those of us left behind were able to acclimatise in more relaxed surroundings. Tropical routine meant we would parade at 0600hrs for a hard session of heavy physical training before gun drills and lectures on jungle warfare. By 1400hrs we were pushing out the strokes in the camp pool, and by 1600hrs we were in the Battery bar trying to come to terms with the variety of taste that Mr Belikin could offer. He offered us plenty, and we took the opportunity to get to know him and Mr Heineken very well indeed.

To be a rounded soldier, in particular a commando, one must be exposed to operate in a variety of environments and climates. As part of a spearhead organisation, one can be mobilised at a moment's notice to any part of the globe; whether operating in the mountains, desert; through winter or summer, a commando must be able to overcome not just the enemy, but the hostilities that the natural world can throw at him. The jungle is an environment that tests the soldier's skill to the maximum, no surprise then that special forces selection regard the six week jungle phase as the most crucial for success, cutting the wheat from the chaff.

The lads returning from the first jungle course looked like they'd spent two weeks on a diet of botulism. All had lost immense amounts of weight. They smelled like roadkill, and looked dirtier than a street corner bin. They'd all enjoyed the experience, but had found it energy sapping, so I was quite apprehensive about proving myself in such a new domain.

Within the first couple of days in 'the trees' I found myself at home. While to some, the jungle is a sinister shroud of shadowed screams and hidden whispers; it was an environment in which I immediately felt comfortable. The sticky heat hardly bothered me, the sunscreening canopy usually making conditions bearable. I found navigation through the dense vegetation, although a challenge, not too difficult, and the movement and patrolling through the jungle I regarded as how real soldiering should be. After the boredom of blackshod training in the UK, I found my appetite for soldiering refreshed. The only thing of worry was the creepy crawlies. Snakes I was quite happy with, despite Belize being home to one of the world's most aggressive snakes: The Combat 18 of serpents, the deadly Fleur De Lance was prevalent in these parts. Being extremely territorial you only had to walk in a Fleur De Lance's vicinity and it would attack. Filling the victim with haemotoxic venom, it would kill within 24 hours without any medical attention.

Spiders, on the other hand, I've an aversion to. Not quite phobic to a disposition of paralysing fear, I'd just rather stay clear of them, a little like I am with religious zealots. Tarantulas were a particular fear. Hearing tales of men waking up in the night with a tarantula laying inches above their head resting on the mosquito net basking in the warmth of exhaled air, I slept on my side above the circles of talcum powder deposited on the jungle floor to prevent my A-frame bed/shelter being taken over by bullet ants - so called as their bite is akin to being shot by a bullet.

Belize is a 'dirty' jungle. It is wild and wonderfully messy, like a nymphomaniac's bed hair. One is almost permanently in contact with the flora and fauna. Ant covered rattan vines strangle the body like an anaconda, making movement hard going. Patrolling through dense greenery promotes the chances of being scratched, bitten, or stung. It also made stealth far more difficult.

What was surprising was that my first loss against the natural world came in the form of a tree. The Latin name is Zanthoxylum. The British Forces name is 'Bastard Tree'. Sporting syringe like needles, when accidentally grabbed or brushed against, the victim would naturally curse 'Bastard.' I did curse when pushing past one but personally think it should be renamed the 'Ooh Ya Fucker Tree'. It was sharp, but like any prick, was only a momentary pain. What was of concern was the ever-growing numbness. It grew until my forearm fell away, unable to grip or stay up. Muscle control evaporated until it eventually hung limply like a prosthetic arm. Not yet panicked, probably because it wasn't my wanking arm, I rolled up my sleeve. Lodged deep into my skin I found two 5cm needles. The medic removed them with his over-used tweezers. Amazingly, the numbness wore off almost immediately and my arm became mobile again in no time. The two puncture marks still leave a scar years later - all quite bizarre.

The jungle did take other casualties. Forty ticks had converged on one lad's back. I watched intently as the medic carried out the necessary surgery to remove them. Pete, the Bootneck I'd shared a train to Arbroath with, returned from jungle school with a blistered patch on his cheek. An animal of unknown origin had apparently bitten him. He thought it may have been a spider, which made me feel fantastic. I couldn't wait to meet my first face-biting arachnid.

An important part of jungle training was the survival aspect. In comparison to other environments, the jungle is a veritable banquet for anyone unfortunate enough to be left in a survival situation. The main problem is that there is as much that will kill you as keep you alive, so it was vitally important to distinguish between edible and poisonous. What was definitely edible was the petting zoo the local instructor brought in to murder for our dinner. An iguana, a couple of skanky chickens, a snake and a turtle were all beheaded for our viewing pleasure. This was no place for vegans.

The killing of the turtle was the most tragic. Trying to lure the turtle's head from the shell, the instructor held lettuce leaves in front of his little head. The turtle would attempt to take the bait but avoid the makeshift lasso that the instructor tried to throttle him with. It became comical, the turtle retreating into his shell, evading capture, after taking a chunk of lettuce, and we cheered every time it took a successful bite. His luck could only last so long. Ever more frustrated, the instructor finally lassoed the neck and pulled tight. The turtle's tongue fell out in evident anguish. As if intent on making the animal suffer for embarrassing him in front of the class, instead of just swiftly chopping off its head, he started to slowly slice through as if carving a Sunday roast. The poor turtle's eyes winced, as we did, until the head pulled away from the tension of the tight lasso. It was cruel and highly unnecessary. From that day forward I vowed never to eat turtle again. I say from that day forward. It was a shame to let the poor little bugger suffer for nothing, so we ate him. To be honest it's a bit overrated, although its unfertilised eggs were quite a delicacy, but turned most off, which meant more for me. As for the other animals, everything tasted like chicken, apart from the chicken. It tasted like shite.

Once everyone was fully jungle trained, we could get on with operations. To ensure we had a good understanding of Belizean culture, and its historical importance within the empire, we took our first free weekend on Caye Caulker - an unspoilt coral atoll that had no cars and no roads; just a couple of simple hotels with plenty of rum.

I couldn't stop smiling. Here I was, basking under the Caribbean sun, on a glass bottomed boat topping up the tan with a few of the lads scuppering unpredict-able Belikin at will, talking about how soft the sand, how beautiful the women and how cheap the lobster, all while watching irradiant tropical fish swim under us. I was getting paid to do this. Life couldn't get any better.

'Don't take yourself too seriously.'

~ Bob Hoskins, Actor

MOST BOOTNECKS HAD now jumped with both feet into Battery life, integrating ourselves so deeply, we nearly called a wet a brew.

I'd been put on the recce party, the other bootnecks placed on the guns. It looked like too much hard graft, that gun-lifting lark, so I was happy being up front, caching out sights for possible gun lines.

Unfortunately, a couple of the bootnecks, Pete and Greg had been somewhat aloof, which all had noted. It was decided by a couple of Battery lads on their gun team to give them a midnight shake to test their resolve. Most bootnecks were torn between the loyalty to other bootnecks and the reality that the two lads weren't really upholding the best traditions of the Corps, so told the lads to prepare themselves for shit but we would stand by them. As the Battery lads were as fair as a Scottish albino, they were happy to just give out a 'gypsy's warning'.

'Why am I getting shit?' Pete asked me while we sat around the breakfast table with a few of the more vocal lads. It was pretty brave of him to front up to an audience of lads that smouldered resentment.

'Everyone thinks you and Greg are anti pongo,' I replied, like some sort of UN intermediary, sipping on my wet.

'Really? Why?' Pete was genuinely puzzled.

'Well you've not really pissed up, you've not talked to any of the lads outside of work and we've never seen you in the bar.'

'Have you possibly thought that I may be just trying to settle in slowly? We don't all have to piss up constantly and nosh each other off to be mates.'

We were all rather speechless. He was right. It had never occurred to any of us that not lashing out the chews and permanent insobriety wasn't necessarily a sign of ignorance. Once we'd realised his intent, Pete was accepted. After a few conversations with him I found him the funniest man alive.

Drunken nights in the Battery bar became the norm. Even on a Tuesday morning we would parade bleary eyed from the 'Mad Monday' we had purposely highlighted as a night for childish frolics. This could range from parading with our cocks out with the cavalry section, much to the annoyance of their officers, or parading in ladies underwear with our own, such as when parading for swimming circuits with the excuse that we 'couldn't find our trunks'.

On the more popular nights, the bar would be full. It would not be an exaggeration to say that we had more nights drinking naked than we did clothed. We would have impromptu naked kangaroo courts, anyone found guilty of any misdemeanour, usually the non-commando ranks attached from other artillery units, would be hung over the crossbar supporting the atap roof and beaten with a fire beater, reaffirmation of the military's caste system where those who had undergone the harder training were at the top, sliding down to those corps where tying shoelaces was a three weeks course. Other than winning the sack race at the 1981 inter-school sports carnival, I'd never really won anything, so to be voted 'Chief Bottom Spanker' was quite an accolade. It would be my responsibility to beat any miscreant until his arse was redder than a baboon's. Fire extinguisher enemas were never regarded as a punishment, on the contrary, and the lobbing of beer cans became the language through which we would communicate, directing an empty one at someone could be translated as a placid rebuke, yet anything requiring aggressive reinforcement of a point would see a full one thrown at great knots at someone's head. The Battery lads worked hard and played even harder.

I did find the army lads more aggressive in their drinking. Maybe it was an army thing, but I was never surprised when the Battery lads kicked off. The infantry lads of the Royal Highland Fusiliers were often targeted, usually for just gobbing off in some incoherent language. While some of the lads were gleaming, many were pasty-faced weaklings, doing little to disprove the maligned theory that the infantry was made up of members of society who couldn't pass the tests to get into any other part of the army.

In between long bouts of heavy drinking we did actually work. I was given the opportunity to work with the Observation Post (OP) teams, made up from the cream of Battery blokes. They were the eyes and ears of the gun lines reporting from far in advance to call down firepower on the enemy. These lads had the highest standards and only if I made the cut would I be able to stay on the OPs. When not out on energy sapping week-long jungle patrols, where hard routines of cold food, silence and laying on the moving jungle floor were the norm, the working week was of continual training on the kit and equipment, mornings of map reading and upkeep of OP skills. The elevated positioning of the OP teams also offered us awe-inspiring vistas, especially at first light where the jungle mist hangs motionless in a cushion of cloud air as if waiting comfortably in expectation of the morning jungle chorus. It would be fair to say searching for rare wildlife was far more exciting than watching out for a Guatemalan enemy that, in truth, would never show.

***

Rather than rest, the weekend for me was always to catch the early morning four tonne truck to Belize City. Ambergris Caye, and more specifically the resort of San Pedro, became our destination of choice.

The ride out to the cayes in itself was a journey of exhilaration. The water so shallow, we could have probably walked the whole way. We sped through the crystal waters cutting a frothing wake over the shallow coral, marvelling at the small cayes that dotted the way, often only a leaning palm tree surrounded by a salt white beach. We expected to see a cartoon character sitting under the palm with SOS written out in stones. We were in paradise and my mind was as clear as the sea it now skimmed over.

San Pedro, had a choice of upmarket hotels, all of which we ignored in favour of the 'Highwayman's Hotel'. Cheap and clean, we could drink our own beers and the dorms would only flood if someone had dropped a log too big for the toilet to handle. Which, if honest, was rare - our choice of drinks was enough to keep us all rather loose.

Time on the caye was spent sunbathing, snorkeling, windsurfing and generally being holidaymakers all at the taxpayers' expense. I couldn't keep away from the place. Belize was drawing me in. Some of the more sensible guys with responsibilities back home rarely ventured out. They were either skint or saving for a new car. I couldn't understand why, while immersed within the brilliance of life, they would choose to see joy only through the lives of others. In my mind a car would be long gone in ten years time, rusted and sold on to someone who would buy it for their son/daughter as a first car or some pikey to thrash around a council estate. Either way, it would be discarded by a once proud owner and forgotten in all but a list of cars they ever owned. I intended to live by my motto of 'You Can't Buy Your Memories', and these would last me a lifetime. In future years when my time comes to smell of piss and biscuits, I will inwardly smile at the predicaments I got myself into in my youth. Here, where the sun met idiocy, I was intent on making the most of my time.

The 'Tackle Box' on San Pedro became our usual daytime hang out. At the end of a pier, the bar overlooked a small man-made aquarium containing nurse sharks, barracuda, a couple of snapping turtles and other less recognisable sea creatures. It seemed a little constrained for my liking, and as the beer took over I hatched a cunning plan.

'I've hatched a cunning plan,' I said drunkenly to Basher, a toothless bombardier, just to reinforce the point of hatching a cunning plan.

'Beer + hatching a cunning plan = foolishness,' he replied.

'But they're usually the most cunning,' I burbled.

His look told me it wasn't necessarily true, so I plied him with more tequila as I explained.

'That's a shit plan, we may get eaten, drowned or shot,' said Basher drunkenly. 'Let's do it.'

It was the dead of night. Under the torchlight of a half moon, the sound of palm trees being stroked by the cooling breeze and the sea swaddling the soft sand calmed any nerves that a drunken fool would have prior to climbing along the underside of a long, rickety pier. 'Operation Release Sea Creatures' was underway.

Looking like week 1 recruits for the Animal Liberation Front, we inched along the slimy, barnacle-encrusted beams until we found ourselves at the edge of the Tackle Box's aquarium. The problem of planning this raid when inebriated meant I'd fallen short on how we were to actually get the creatures into the sea. A bit of a conundrum it became so into the aquarium water I went, totally oblivious to the fact that snapping turtles were called so for a reason, sharks were not to be trusted, and barracuda in shoals would rip a man apart like piranhas on strong lager. Looking for an escape hatch, Basher hoarsely whispered, 'There's someone coming.'

He was, of course, quite correct. What he failed to mention was one of them had a long pole, the other a shotgun.

With little in the way of salutary greeting, the man holding a shotgun said, 'You got two options. Get out or get shot.'

It didn't take me much persuading to get out of the water, nor did it take much for me to wince when the pole was whipped across my back. A little upset that this pole attack hadn't been part of the options offered, I nonetheless jumped over the side into the sea to swim along the coast scrambling onto the beach giggling at our patheticness.

The following afternoon, showing off my huge red welt as a badge of dishonor, we snorkelled around the next jetty along from our failed raid and discovered a shark still in the shade of the pier. Sharks permanently swim and never stop when alive, so by this poorly judged conclusion of rudimentary marine science we deduced that the aforementioned shark was dead. Bally, Pete and I agreed that we should capture it, drag it onto land and take photographs claiming we were the greatest bootneck shark hunters ever to walk upon God's earth. Not that we had any fishing gear. The nearest scavenge of anything fishing-like uncovered a bamboo pole and some discarded fishing hooks. Using the fishing twine already fastened to the hook, Pete lashed it to the bamboo pole. We now had a hooking implement that was incapable of hooking a stickleback never mind a 200kg shark. Ignoring the laws of physics, we continued in our quest to drag up a shark. Allocated the job of photographer as I owned an underwater camera, the catch would be captured for posterity. Bally, took the role of hooker, Pete would assist the lift once Bally raised the shark from the sea bed.

Bally swam above the shark, and poked the hook as if trying to catch a fairground duck in the vain hope of winning a cuddly toy that would fail any modern safety standard. Poke as he may, nothing caught, so he tried more of a podiatrist's scraping method hoping to catch the thick skin of the shark before pulling it up. He scraped once, then twice. On the third scrape he must have caught the skin, not that it lifted up the shark, only anger it enough to swim up to defend itself.

I took lots of photos of bubbles and pier legs as we swam for our lives. Walking on water like cartoon characters, we tried to escape the shark who was busy trying to catch whichever idiot was stupid enough to strike its back with a rubbish fishing implement. Unless it was 'Shark Jesus' it appeared it hadn't died.

Upon telling our stories to a local later, it appeared that nurse sharks were very common in these waters and were one of the very few shark species that can rest immobile, often found laying still in the shade of the jetties. Well, you live and learn.

Yet even here, where I found sanctuary in stupidity, for no apparent reason, my mood would again turn to surliness. There was no rhyme or reason. I wanted nothing more than to be left alone and my Walkman became my mood therapist allowing me to wallow under the pretense of listening to music. I'd perfected my excuses by now. Heavy alcohol intake meant an excuse of being hungover or just knackered covered my mental tracks and should my low feelings persist beyond the night I sat quietly with the OP lads offering to do any undesirable chore that allowed solitude.

Once the darkness dispersed, pushing the barriers of normality became a personal quest, as it did for a few others. If there was lunacy to be had, there we were at the front of the queue, hand high in the air, volunteering for any shade of nonsense that may cause embarrassment, hilarity or long term injury. Whether it was surfing on top of stolen golf carts and being jettisoned into a mangrove swamp, diving off 50 foot high rocks head first into unknown river waters, or inventing new dances at 'The Blue Angel' nightclub such as the 'kickstart dance' (dance on one foot kicking your other leg like starting a motorbike - great for knee problems) or the 'orange dance' (dance and stare at an imaginary orange held above the head - great for looking a tit); I distanced myself from those not up for permanent fun. At each and every opportunity I'd be out of camp looking for adventures new. Belize is a treasure trove of activities and after each working day, if I could organise transport, I'd be out exploring. We were only here for six months, so had to be out and about making the most of this privileged opportunity. Whether it be visiting thousand foot waterfalls, sliding along natural water slides in the leech infested Rio On, or just shopping at the local markets in nearby San Ignacio buying bits of shit I'd never use unless I metamorphosed into a middle aged Belizean mother, Belize made me feel alive.

While I was having the time of my life, my girl-friend, Toni wasn't. This was confirmed to me when I received a letter. The previous correspondence had been laboured, not the usually lighthearted affair. This particular letter confirmed the worst. It was the dreaded 'Dear John'. She was binning me. We'd been together around six months but had only spent around three weeks together. What relationship there was, was over.

I thought I better call her to see what the situation was. A bit late really, we'd been away for nearly three months and I hadn't yet phoned her. No wonder she claimed I was putting the marines before her. Receiving a 'Dear John' meant two things: I got beer bought for me all night, and secondly I was now free to commit whatever acts of depravity my salacious mind could drum up.

After only two hours from pinning my 'Dear John' on the wall in the bar, I found myself drunk as a skunk in the local whorehouse. Four Battery boys and Pete accompanied me. While some entertained themselves in the company of Guatemalan ladies, I sat on a bar stool sipping Belizean rum with a lady of the night who rubbed my leg for so long I feared friction burns. I say lady 'of the night' as I could not envisage her being allowed out during the day. I can't lie. I wasn't keen. I don't know if it was the sawdust on the floor but she smelled of caged mice so her chances of getting me to fork out my easily earned money for fornication were absolute zero. Pete was happy enough, he was receiving plaudits for being rather proficient at dominoes, beating the pimps and prostitutes repeatedly with his bold and offensive double hook moves.

The downside of being dumped was that my R'n'R plans were now in disarray. I'd planned to meet Toni in the Dominican Republic for a fortnight of love-themed activities, such walking along the beaches at sunset, romantic dinners under the stars, and arguing about watching football in the hotel bar.

I was now stumped. My closest mates had organised their leave and mine didn't coincide with any of them. Lee, my ridiculous dancing sidekick, suggested I join him climbing in upstate New York on adventure training. Requesting whether I could tag onto this climbing trip I was told by the Troop Officer that yes, I could go adventure training. 'But you realise that adventure training will be your R'n'R as well.' It was a statement of confirmation rather than a question of understanding.

'Yes, Sir,' I replied sincerely.

'Good, so no trying to pull a fast one and getting more leave.'

'No, Sir.' Sincerity was still imbued within. I had no alternative plans, rock climbing in the States for two weeks seemed a fine way to spend my holidays.

Before any holidays could commence though, much soldiering was to be done and within the OP team to which I was attached, the soldiering was harsh.

The OP team signaler was Mick Smith. Mick was a commando in the truest sense of the word - tough, gregarious, a larger than life professional soldier, and a terrible dancer. When I met him he was already well known throughout the regiment. We got on immediately. Of similar age we shared a similar sense of humour and the same taste in music often swapping tapes to widen our libraries. We would bump into each other from time to time throughout our later careers; invariably he would be in the shit but always climbing up the promotion ladder. Just weeks before he was due to retire from the Army after over 22 years sterling service, his life was tragically cut short, killed by an enemy grenade while serving with distinction as a warrant officer in Afghanistan. Mick left us, yet is immortalised as a 29 Commando legend.

My everlasting memory of Mick in the field was the beautifully yellow huge boil-like zit that had developed under his ever-shorn head. As well as painful it was becoming an encumbrance and also the subject of many jokes. The un-squeezable zit was named Colin, and so big it could have worn its own jungle hat. Finally, the medic decided the spot needed to be sorted. With his field surgical kit in hand we crowded around the top of Mick's head to watch the ultimate zit being popped. Unable to squeeze it by hand, the medic decided cutting a 'x' across the top of the zit would relieve enough pressure for the rest to be squeezed out. How right he was. As soon as the incision was made, foul smelling yellow pus ejected high into the air. It was thoroughly disgusting but we all watched with morbid intensity as the medic tried to squeeze out the core. This deeper pus was thick, like clotted cream and blood appeared until finally, 'pop', out came the core. Only it wasn't a core. It was 2" long worm that had burrowed itself into Mick's head and spent weeks feasting on his scalp. Nice.

***

It was the summer of 1990. England was playing in the FIFA World Cup in Italy. 'World In Motion' by New Order overtook the Stone Roses as the music of choice and played through the humid air from every English-speaking radio station. We were in Central America, but our cheers could be heard as if we were in Bari itself.

Our haunt for the World Cup was the Red Rooster bar in San Ignacio. Owned by an American couple, they were bemused at our love of 'soccer' but it filled their coffers as the tournament started, the bar packed to the cobwebbed rafters during every match. Knowing the lack of motivation of the troops while important games played out, we were kindly given the mornings off for both the England and Scotland games. Of course, I supported England and cheered on the Scots when they played, after all no one cheered louder than I when Archie Gemmill scored that goal against Holland in 1978. The Royal Highland Fusiliers supported Scotland and anyone who played England. It was a disappointing affair for the Scots. While they beat Sweden they fell against the powerhouses of Brazil and Costa Rica, the loss against the latter saw us Sassenachs give reason to have a poke at the very unhappy Jocks. England, for once, surpassed expectation, and as we progressed through the tournament the excitement and anticipation grew. Some of the Royal Engineer unit that was with us decided the World Cup was a good excuse to 'mug a Mennonite'.

The Mennonites are a strange, but harmless, group of Dutch immigrants who have abandoned modernity in favour of a pastoral existence. They had a sizeable community in Spanish Lookout, a small settlement across the river from Holdfast Camp. Luddites in every sense, they hand farmed the land preferring ancient techniques over advanced methods. Accepted by the Belizeans and largely ignored by us, our paths rarely crossed apart from the odd time on troop runs where we would have to make way for their carts.

The Royal Engineer hooligans had different ideas. England was pitted against the Netherlands in the group games. As hooligans often do, they targeted anyone or anything associated with their opponents. Unfortunately for a pair of Mennonites, due to their historical Dutch links, they became victims. Some drunken squaddies attacked a couple of Mennonites while prowling the streets of San Ignacio. While it never got reported, there was little support for those culpable. In the second round England played Belgium, a rather dour game ending with David Platt scoring a last minute winner. Whether the original Mennonite attackers were responsible, it was never discovered, but again some Mennonites took the wrath of football's most geographically challenged supporters who deduced Belgians, bordering the Dutch, were apparently from the same country.

It was July 4th - US Independence day. The owners of the Red Rooster kindly organised a barbeque to celebrate the occasion upon the conclusion of the Semi final between England and Germany. Germany, sharing a border with both the Netherlands and Belgium suggested another attack on the Mennonites was afoot.

While they may have tilled the soil with 18th century tools and grown beards without moustaches, they were acutely aware of events in Italy. They were also aware of the pattern that had emerged during the World Cup tournament. They entered the Red Rooster bar and informed us that should there be any attack on their own from any British soldier then they would be shot. Like their existence the warning was simple. They may not have tractors, but they had shotguns. The warning was heeded. No one mugged a Mennonite after that.

*'Soldiers can sometimes make decisions that are smarter
than the orders they've been given.'*

~ Ender's Game, Orson Scott Card

WITH ADVENTURE TRAINING approaching, I
was asked to take part in a charity run from Rideau
Camp in the South to Airport Camp the British Forces
HQ on the outskirts of Belize City.

It was an idea spawned the previous year by the
lads of the Parachute Regiment - now there's a first - to
try and raise money for a para sergeant's child suffering
a debilitating illness. They'd chosen the route, 213 miles
of arduous terrain, in a team of ten runners and had
completed the course in just over 22 hours. It was
hoped the run would be kept alive by the following
units and we decided we should continue tradition. Not
only would it be for a great cause, but also we had to
beat the paras.

Ten of us were chosen. I can only think I got the
nod due to them wanting someone to do impressions
with their genitals in the back of a vehicle - my 'goose
flying south for the winter' only slightly behind my
'Angel Gabriel' as a 7 Battery favourite. It took us the
best part of the day for us Holdfast Camp lads to get
down to Rideau Camp in the south of Belize. We
arrived early evening and were fed and watered while
meeting the other runners based at Rideau. For the
Battery lads, it became an impromptu reunion after a
few months separation; for me it was good to meet
future mates. For the sake of sensibility prior to an epic
run at 0400 the next morning, we just had a solitary
beer and went to bed around 2200hrs.

'Why you in bed, ya bunch of lightweights?' asked one of the Rideau runners. It was a nice way to be awoken.

'We're running at four,' replied someone post-slumber.

'Four in the afternoon, ya dick. We're running through the night to escape the heat of the day,' said the Rideau lad.

'So much for 24 hour military time keeping then,' I said.

Thinking it would be unsociable to not join the lads in their bar, we had another beer, and another and another. We left the bar at 4am, totally mingbats. Only 12 hours to sober up before we were to each run 21 miles. Well at least it wasn't a marathon.

Waking in the early afternoon with a mouth like the bottom of a birdcage, I joined the lads to collect the stores for the run. The camp quartermaster was a paratrooper on an extended deployment and had organised our rations for the run. He must have had a wager that we wouldn't beat the paras' time and also his regimental pride was evident as the rations issued were a shit load of water, a sack of potatoes and a sack of carrots. Raw. With no cooking facilities and our beds the floor of the four-tonne support truck, rather than whinging like bitches, we saw it as an ideal opportunity to get one over on the paras. We'd take the rations so when their time was beaten, we could pride ourselves that success had been gained the hard way, on raw vegetables and insomnia, oh, and the $50 worth of chocolate bars that we bought from the NAAFI.

The fastest way of completing the distance, we figured, would be to run in a rotation of two-mile legs, thrashing ourselves for that short distance before resting until it was our turn again.

We set off into the jungle-lined dirt roads that led from Rideau camp. Upon the sun departing, we fell into a natural night routine of sleeping and eating. Every two hours or so, we'd be woken from our restless bone-jarring sleep, as if on sentry duty to ready ourselves for the next leg. We'd jump into the supporting Land Rover to wake up properly while the man in front finished his two miles. At the correct interval we'd jump out of the vehicle and carry on at the target six minute mile pace.

There is little better than being awake at night in a tropical country. The tropics excite the nostrils, redolent with unique exotica, and nighttime adds a quirky potpourri to the odorous air. Even running up hills and along potholed dirt tracks could not take away the exhilaration of being here. In the middle of nowhere running freely, with vehicle headlights the torches to breach the darkness ahead, strange animals hid behind the curtain of jungle foliage, their position given away by the bright reflection from their inquisitive eyes. It could be as benign as a skunk or as grandiose as a jaguar. Whatever watched, it only enhanced my joy of running.

If the nighttime was enjoyable, running during the daytime heat was harsh. Having each covered plenty of miles, the stretches became more painful for our increasingly heavy legs. Another vehicle from Airport Camp joined us pulling a bowser full of water - a cooling dunker after each scorching leg. While there were stronger runners, there was no individual glory to the run. Together we ran the last mile as a team, finishing the course in 20 hours 33 minutes, beating the paras' time by nearly 90 minutes.

Already on our knees, we were smacked around the head with a large dose of irony through the lack of recognition of our charitable and physical success. Not one person at Airport camp welcomed us - everyone was too busy watching overpaid English footballers losing the third/fourth play off against Italy at the World Cup.

***

With Upstate New York on my mind, I packed for two weeks climbing. My climbing gear consisted of gaudy lycra shorts and a bottle full of 'courage' pills. The few cliffs I'd climbed were pretty basic and nothing compared to what we were to face. I wasn't overly keen on hanging off a smooth rock face with only my pinky finger wedged in a crack the width of a mouse's ear preventing me from falling to my untimely death.

The climbing expedition met at Airport Camp (APC). The lads from Holdfast met up with the lads from Rideau and again new acquaintances were made. I immediately hit it off with a Battery lad from Rideau called Bungy, and it was decided a communal night out would be in order to foster our burgeoning friendships. The obvious place to go wasn't the NAAFI, but the infamous brothel 'Raoul's Rose Garden' that made its fortune from APC that sat across from its flaking front doors.

We'd been told that Raoul's was strictly out of bounds to APC personnel as there were too many servicemen reporting sick with syphilis, gonorrhea and other nasty ailments attributed to putting their bits into unsafe places. However, we weren't APC personnel, and those with a dose obviously weren't commandos. A green beret evidently gives more protection than a condom. So, with strict orders totally ignored, we jumped across to Raoul's to partake in some refreshment. The ladies with the bear trap vaginas were more than happy to see us, and despite the ban it was clear others had defied their bosses orders as a sprinkling of servicemen sat in dark corners. The dim (or broken) lighting only added to the seediness, and behind the huge bar was a room with a corridor that divided the rows of curtained booths. Behind these heavy red curtains lay a thin mattress placed on a low bed next to a box of tissues. It was battery farm prostitution and about as erotic as stepping in dog shit, but was where many of the squaddies spent their money.

With all the surprise of a charging elephant, the doors smashed open. In bolted half a dozen Military Police with nothing better to do of an evening than catch out anyone tempted to dip their wick in a dirty cervix. Sitting at the bar gave ample opportunity for Lee and me to hide. Jumping like chased gazelles over the bar we sat at the feet of the barmaid who giggled at our signals for her to be quiet. She not only agreed, but gave us each a bottle of Belikin beer to keep us occupied while the Redcaps went through the club. We could hear the pathetic excuses of those caught and also from some of our lads who remonstrated that they were visiting and didn't know there was a ban. With plenty of guilty bastards already caught, the Redcaps accepted their excuse but threw them out warning them not to return. It left just Lee and I in the company of about 20 working girls who loved the fact we'd shown true commando guile in not being captured. With a ration of ten prostitutes to one man you'd think we'd succumb to carnal pleasure, yet gladly left with tamper free penises.

The expedition leader, Lt C was a meek, yet stuffy chap. With a public school education carved into him, he was about as far removed from the lads as one could possibly get. Officers were the biggest difference I found with 29Cdo and the Corps. As Royal Marine officers train on the same camp as recruits, there becomes an automatic kinship that seems rare in normal military circles. Some of the best Royal Marines officers have been as rough as guts, some even displaying home made tattoos. This is not to say they don't lead as an officer should, but they command respect from their men while immersing themselves in the culture of a bootneck. I never found this deferential camaraderie with any of the other arms I ever worked with. There always seemed a more divisive line and Lt C was a man who liked to keep a big thick line between him and his charges. He'd already laid down the law even before we touched down in Washington DC. We were on adventure training not a jolly, and we'd be living quite rough under canvas to gain optimal time on the cliff face. Just how rough I wasn't sure, but I was pretty adamant I hadn't signed up to living in a forest. His early briefing did give us time to plot a counter offensive. No one was particularly keen on living in the wilderness sharing the shithouse with a bear.

The drive from Washington DC to New York State is a long one when cramped in a van full of overgrown gunners and expedition equipment. As we headed North up Interstate 87 freeway the tree-lined beauty of New York State surrounded us. We saw signs for wonderfully named places like Poughkeepsie and a familiar place that rekindled bad memories - Woodbury, where we'd all spent countless uncomfortable hours.

We were headed to the small town of New Paltz and arriving there early morning we felt it a little British. We'd seen many British named towns on the journey and here we saw a Barclays bank, and discovered we were in the middle of Ulster County. What was also a surprise, but a very nice one, was that New Paltz was the home to the State University of New York. This snippet bore no change in Lt C, so following him into the nearby forests we searched for somewhere he thought suitable. With it came the realisation he was intent on living as basic as possible, he wouldn't even consider a campsite. The foot of a cliff was, in his mind, the perfect spot to sleep. It would be like setting up a harbor position. Our counter strike needed to be put into operation.

When you have a university town nearby why would you honestly sleep in the woods with only a babbling brook for company? With nearby sounds of American teenagers screaming over the top of disco music, there could not be a lonelier place than under a tent sheet at the bottom of a cliff face. We pressed Lt C to reconsider. It was only at this point that he admitted the difficulty in going elsewhere. He didn't have sufficient funds to get proper accommodation. Here he was, the highest paid member of the expedition, with the least money. He was taking the word 'adventure' far too seriously.

'But boss, a campsite is not going to cost much between the lot of us a fiver each at most,' said Bungy.

'I can't afford that,' replied Lt C.

'A fiver? How much did you bring?'

Lt C looked sheepish. '$50.'

We all looked at each wondering who was going to burst into laughter first.

OK, he was a graduate but now he was an army officer, therefore hardly necessary to regress back to a student and live off instant noodles hydrated from boiled stream water. We didn't want five star hospitality, we just required somewhere we could wipe our arses with something other than a dock leaf.

Bungy and I became the recce party. We spotted a campsite that looked perfect. Indeed it was. While they couldn't accommodate us with pitches they had an even better solution to our accommodation needs - a spare barn that was clean, had running hot water and a toilet. What more could we want? It even had a broken pool table. Even Lt C couldn't complain. But he did, and his continual protestations became nothing more than a childish whine.

My first task, once settled, was to forge some fake ID. Still only 20 years old, I was three months short of being able to drink legally in the US. With the cunning of a fox who'd failed cunning studies and the artisan skills of a myopic lobster, I managed to fashion my real ID card with a slice of a knife to delete a '1' from my date of birth. From my birthday being 13.11.69 I was now born on the 13.1.69. It couldn't fail.

By pure coincidence, the local university campus was on its fresher week where all the newbie students are initiated into university life by the introduction of underage drinking and promiscuity competitions. For a group of young British males who, for the previous months, had only woken up to the sight and smell of hairy arses poking out of bunks, the thought of waking up next to a sweet smelling American girl was a little too much to resist. And what a catch we would be? They had four years to meet the right life partner, an intellect who may one day be a senator, a doctor, or a judge. What they needed right now was an irresponsible bootneck wearing Union Jack shorts and an inane ability to talk shite in a strange accent. It was this accent, a fluent mix of West Yorkshire and bollocks, that managed to woo one young lady who, surprisingly, was quite attractive and with a smile that I totally ignored in preference to ogling her mammary glands, willingly took me back to her university dorm to conduct an experiment to see if the male genitalia were the same across the pond. Using me as a laboratory guinea pig, empirical evidence probably concluded that British men had far smaller penises, but could go like a piston engine.

A 20 year old male has many dreams: being able to play for their country at any sport, to have a successful career, own an electric carving knife, have the opportunity milk a cow on a warm summer's morning, or to pepper spray a politician. One of mine was to be woken up by a bevy of beauties dressed in all manner of lingerie. By the miracle of miracles this actually happened. After being made to feel cheap by being asked to sleep on a couch in a communal room, I was awoken by her student friends who were all getting up for morning classes. Apparently I was a bit of a celebrity, nothing to do with my sexual prowess, but my accent. So before their lectures they needed some tutoring on how to speak proper England like I did when I were a little children.

I felt like a circus freak being asked to say words they already knew but slightly differently. What do you say when asked to say something? 'Hello,' sounded meek. 'Big fat elephant's fanny,' would sound downright weird, so I settled with, 'Anyone else want a bag off?'

Should they have understood the connotation, I am sure I would have been thrown out.

'A bag of what?' asked by one delectable nerd, suggested they didn't. Yet just saying these few words made me seem the funniest human in existence. Trying to avoid an encore, I excused myself for a pee 'accidentally' walking into the female toilets. Oh those American restrooms.

Not wanting to incur the wrath of Lt C who was still whinging about our luxurious surroundings, I returned in time for our first day climbing. Thinking I may be a weak member, I found, with my puny body, I could climb quite well. With this confidence, my initial trepidation turned to enjoyment as we negotiated the short pitches and the abundant scrambling routes. I even started to lead climb. My technique was positively amateur using pure brute force and ignorance to scale the heights. This will get you so far but when the routes became harder my lack of technique saw me getting stuck in the most precarious positions, much to the hilarity of the spider monkeys, who could climb windowpanes.

The trip to New Paltz was how adventure training should be - the daytime pitting minds and bodies against the physical challenges of climbing, then as the night fell, challenging livers and stomach against alcohol and takeaway junk food. I found the delights of peppermint liquor that tasted like alcoholic mouthwash, gave no hangover and left my mouth feeling fresh as a daisy until I filled it with either a burger or a vagina if I was lucky. After a hard week's climbing, New York City beckoned. While only three of us were tempted to take a bite of the 'Big Apple', we managed to fill the car going back with more shopping than any footballer's wife could manage. Just being in famous department stores or seeing anything emblazoned with 'New York' was enough to part with cash.

Even Lt C had calmed a little. Although a shit budgeter, he was a fantastic climber, wearing scabby trainers to scale routes that the rest of us couldn't climb even wearing specialist 'stickies'. With his climbing thirst satiated, he chilled to a point where he allowed us to partake in other activities of our choice. Whether it be kayaking or just going exploring nearby Woodstock or the surrounding countryside, we managed to make the most of our short time there. Lee managed a return to form by upturning a golf cart in a rather plush golf course. We managed to conclude the trip with few injuries, but bergans bulging with memories.

It had to be said I wasn't doing flick flacks at the thought of returning to work. I'd enjoyed my R'N'R/Adventure training and even though I'd agreed, I felt stupid to have forsaken my holidays. Bungy was heading straight off to Cancun and suggested I join him. I told him of my pact with the Troop Officer, but Bungy was a bit of a golden bollocks with the BSM and reckoned he could get me away for an extra two weeks. Uncertain whether this was a good idea, we rocked up to see the BSM. He was an old school stalwart, with a traditional sense of fairness and a Victorian ability to scare the shit out of anyone who crossed him. Explaining my situation, the BSM pondered for a short while.

'You realise your Troop Officer has warned me about you?' he said eventually. 'Cancelling your leave for adventure training? You must be mad.'

My heart sank.

'However, you are entitled to both and both you shall have.'

I looked at him thinking I'd misheard.

'I like people with balls and you, son, have the neck of a thousand giraffes. I will sort it out with your Troop Commander.'

I stood there as if stuck to the linoleum floor.

'You still here? Go on fuck off before I change my mind. Enjoy yourself and keep out of the shit.'

Within five minutes we were waiting at the bus stop outside Airport Camp awaiting our overnight transport to Chetumal on the Mexican border.

'The one unforgiveable sin is to be boring.'

~ Christopher Hitchens

ASK ANY GLOBETROTTER WHAT separates a traveller from a tourist and many will say it is how they proceed from place to place. A tourist, for instance, will sit on a bus that is comfortable. It may have a steward service and will certainly have a guide at the front reeling off fact after fact pointing out points of interest purposely alternating between a sight on the left and a sight on the right to conduct an orchestra of occupants to look as if they are watching tennis. A tourist will spend the night at a predetermined point with a concierge willing to take their luggage in return for a tip that will be accepted with feigned gratitude, and their meals will be served at an appointed time to an appointed menu that has already been agreed upon.

A traveller will eat street food while they sit on a bus, an open window the sole point of temperature control. They will share their seat with a flatulent stranger who lacks the basics of hygiene. Chickens, goats and the odd child will spill around the seats that shake as their bolts come loose from the rusted floor below. A traveller will be on that bus for a long time, using it as somewhere to sleep to save money on a hotel room.

Bungy and I were travellers.

With a change of bus at Chetumal bus station taking three hours longer than it should, we carried on our journey arriving in Cancun a full fourteen hours after leaving the BSM. As we dipped our toes in the early morning coldness of the Cancun surf watching the relaxed sun rise lazily from the ocean horizon, we knew having to smell the eggy farts of the lard arse on the opposite seat, and watching dry goat shit roll under our feet had been worth it.

We wandered the deserted streets, drinking in the sleepiness of this party town. It didn't take long to find somewhere to stay and met Pete and Bally who'd just arrived on their R'N'R. Frivolity would be at the top of the agenda.

And so it passed, as decreed by the Lord of Tequila, that we spent our waking hours wearing vests and union jack shorts looking as British as John Bull, promoting hedonism as a national sport. If it wasn't beer it was tequila, if it wasn't tequila it was mescal. If it wasn't mescal it would be rubbing alcohol. But when in Mexico, tequila had to be the main source of inebriation. Showing our bootneck skills to the max to the adoration/disgust of the mainly American tourists we taught them how to snort vodka and, more importantly, how to prevent wastage by catching the prevailing sneeze in our open mouths, and the time honoured 'ninja tequila slammers' - snorting salt, downing the tequila and squeezing lime/lemon in our eyes. 'Drink responsibly' advocates, we were not.

Our following was unsurprisingly small, unlike the large Canadian girl who took a shine to my inability to drink from a jug spout and my accidental elbowing of her breasts as I wiped away the mess from the aforementioned pouring. She was here on holiday with her friend who it seemed was in her room having a post-relationship breakdown. Enjoying the evening on her own, she had to return to ensure her friend was OK, but invited me back all the same. When she stood up I realised I would look like a child next to her. She stood well over 6ft tall, with shoulders like an American footballer and each foot could cover a small child.

Asking her occupation I shouldn't have been surprised when she replied she was a lumberjack. With Monty Python ringing in my ears I waddled to her room. If she did want sex I may get harmed. She was too big to handle for a wee lad such as myself, but with courage as one of the qualities of a commando, I steeled myself just in case. I had no need to worry. Her mate was still in breakdown mode, a floral tribute of scrunched tissues scattered around her bed where she sat, her face red either from overdoing the crying or poor application of sun tan lotion. With sympathy not high on my list, I decided to leave.

'You'll meet me tomorrow,' said the lumberjack, as an order not a request.

'Yes of course,' I said.

'Ok, I'll meet you at your hotel reception.'

Giving my location to a stranger, especially a female Kraken, as I would find out, was an amateur mistake.

Feeling as though on death row, I awoke mid morning in a cold sweat. The hotel pool bar was calling me to chill, hoping upon hope that the Kraken's friend would decide to return to Canada to resurrect her relationship. Bungy was down there already and laughed as I told him of my predicament. He was already in conversation with a group of holidaying English girls, so I joined in the conversation eager to take my mind off the possibility of being eaten by the Kraken. Within an hour, I was in bed with a girl from Harrow. It had come as a complete surprise but whatever was in her morning beer turned her into an insatiable vixen. Shortly into our session of bedroom gymnastics, I received a knock on the door.

It was Bungy. 'Your lumberjack girlfriend is downstairs,' he whispered with a smirk.

'You know I'm with this Doris, did you not make an excuse for me?'

'Not my problem to sort, Casanova,' he said. 'I said you'd be down in a second.'

He walked away mimicking digging a very big hole that I'd made for myself. I felt like a character in some Robin Asquith comedy film 'Confessions Of A Stupid Bootneck'.

Excusing myself from the now perplexed Harrow girl, I ran down to the atrium to see the lumberjack. Hopefully, she would shrink through sober eyes. If anything, she'd grown, as if taken George's marvellous medicine.

'Hiya, sorry I never noticed the time, could you give me 20 minutes to get a shower?' I said.

'Yeah OK, no problem,' she replied through a face of disappointment. She was evidently a time-conscious lumberjack.

I should have just told her I wasn't interested. But that would have been downright rude. My inability to turn down a woman meant I had to do the honourable thing and at least take her for lunch. Besides, I'd never bedded two girls in a day. Back up the stairs I ran and jumped on my bed out of breath.

'Who was that?'

'Just the lads trying to organise the day. Now, where were we?'

Just as we were reaching the climax of our sweaty writhing, the room phone rang. It was Bungy.

'You better get down here, the lumberjack is waiting at the lift to come up.'

'How's she know my room number?'

'Well if I hadn't told her she would have become suspicious,' said Bungy, relishing his role in my discomfort.

'Twat. OK I'll be down in a minute.' I replied.

I quickly put on my shorts.

'Where you going now? Don't tell me you're leaving me again.' She was now getting riled. My swollen testes, primed and ready to explode were now starting to ache.

'Be back in a minute. The lads need me to give out a deposit.' Lying was now becoming easy. 'I'll be back in 5 minutes.' I had to tell the Canadian to do one. My two girls in a day fantasy was turning into an admin nightmare.

Covered in sweat, I ran back down the stairs to meet the impatient lumberjack. Not really knowing how to say 'fuck off' nicely, I sat with her, the pregnant pause amplifying my unease.

'Look…' I started.

'…You're sweaty,' she interrupted.

'Smile,' said Bungy with a camera at his face. 'You know, Mark's not stopped talking about you since last night.'

'Really?' smiled the lumberjack putting her arm around my clammy neck.

I would have scowled at Bungy if it weren't for the fact I was trying to un-blind myself from his camera flash.

'Oh get a room you two,' said Bungy stirring the pot even more. 'Actually that's a good idea, why don't you go up to his room?

I stared at Bungy. He was enjoying this. I wasn't.

'Who the fuck is that?' The voice came from the lift entrance. It was the Harrow girl with an unresolved vagina making a beeline for me.

Twenty minutes of low quality, sweaty foreplay does not a relationship make, but I could understand her ire to see me sat with another woman.

'Who's this?' It was a fair question by the lumber-jack.

Not really knowing how to handle such a situation I just sat like a bell-end realising any excuse would sound bollocks. It was the first time I'd ever been labelled an 'alley cat', especially from a Canadian, yet it was quite kind considering the Harrow girl called me a 'horrible little cunt', which, in any English speaking country, is slightly offensive and something I'd not been called since Royal Marines training.

As both girls disappeared, the approaching hotel's mariachi band maintained my spirit. A trumpet player plonked a ridiculously large sombrero on my head and Bungy thrust an ice cold beer into my hand. It was only now that I realised I had my vest on back to front. I looked exactly how I felt - an absolute tool.

***

I am rubbish at jigsaws and golf. Both need a degree of patience that I do not possess. As people close to me will attest, I get bored far too easily and now here I was in Cancun wanting to further my adventures. I'd drunk more than a bull elephant could handle and was fearful of a British or Canadian revenge attack.

On a slightly more upbeat note, my climbing skills had improved no end. After the New York trip, the climbing bug had bitten me. Unfortunately, the bug only sunk its mandible/proboscis/labellum into me after yet another infelicitous affair with booze. If I had been sober at midnight, I should have trotted off to bed. But minging, all I wanted to do was climb. With fear, the second-to-last thing on my mind (one in front of 'sense'), my second floor balcony would become the start point for me to free climb naked up the outside of the building. With the façade made up of large boulders

it wasn't a hard route to traverse around the accommodation block, but without any safety equipment,
moronic climbing of this ilk could see a messy death
only a bad hold away. Amazingly, each night I would
return to my room uninjured, and would wake up in the
morning looking at my scarred hands and toes chastis-
ing myself for being such an idiot. But there I would be
fourteen hours later, naked as the day I was born, just
with a slightly better tan, doing it again.

I'd spent a week in Cancun imitating Caligula but
I'd now done it, seen it and got a cheap T-shirt, it had
become routine. So while trying to take my mind off
the permanent shits I was suffering from after endless
beers and dodgy tortillas; in a moment of opportunistic
impulsiveness, I purchased an air ticket from a back
street travel agent that specialised, I suspected, in one-
way tickets to the USA.

Only when I sat in airport departures did the realisation strike me that the girl I'd been writing to didn't
know I'd be visiting her. I couldn't notify her now, her
phone number was back in Belize somewhere in the
devastation of my locker. I couldn't even get international enquiries to help me, as her address was on the
same piece of paper in the same locker in the same
country that I wasn't currently in. *Shit*.

My ex's sister Marie, was working as an au pair to a
Doctor and his wife in San Francisco. We had been
writing as platonic pen pals, so thought it worth dropping in on her unannounced which, in retrospect, was a
little weird.

The last thing Marie expected to see was me. So it
was with shock equal to finding Adolf Hitler as the
gardener that she welcomed me, totally flabbergasted
that I'd make such a journey just for her. I'd managed
to find her by remembering the street name and

knowing that its number was a combination of an 8, 4 and a 5. The Doctor's wife was even more surprised and despite the fact a strange man had turned up out of the blue she allowed me to stay in the cellar. This was special American hospitality, and was most appreciated especially as I was pretty skint, with my end of the month pay still a few days away.

I was fortunate that Marie had a couple of days off, so she showed me the sights and sounds of San Francisco. I was back to being a tourist and where money allowed, visited the well known spots with my camera to take 'this is me in front of...' pictures that would bore people to death within 30 seconds. Even having spent only a week there it left me in no doubt that San Francisco was my favourite city in the US. I loved the bohemian vibe and the liberal attitudes of the majority I met, which was a bonus as it was probably the only reason I didn't get arrested for pissing in a doorway at 2am dressed in nothing but a green plastic Teenage Mutant Ninja Turtle outfit sized for 12 year olds.

I found I was a little more skint than I thought when I flew back to Cancun. Attempting to draw some money from my account I found I was over my overdraft limit. In my pocket I had about $20 to last me the next 24 hours. I had no hotel booked but at least had the return bus ticket back to Belize. I probably should have tried to get the bus sooner but I wanted to squeeze every last inch from my trip. I found being alone in such a wonderful crazy place as Cancun allowed me to see the things I wouldn't normally see. I wandered the streets eating the cheapest street food available and sauntered along the beach, venturing further than I'd done on the previous visit. Lined with luxury hotels its low-rise skyline seemed to ooze American Costa Del Sol. Indeed, when with the lads, I saw more tourists than locals, the younger American crowd flocking down here to escape the stringent drinking laws of home. It was party town central and a place that I'd enjoyed tremendously but one I sensed I'd missed a great deal of due to my activities dictated by the nearest available drink. I looked around at the activities on offer. Paragliding, cultural trips, windsurfing, jet boating, studies on Mayan ruins, snorkeling off the nearby Cozumel coast; I'd missed it all. In truth, it would have been hard for me to afford the majority of activities, but here I was now totally brassic, with no alternative but to wander the streets aimlessly until the bus departure the following morning.

Sleeping rough is usually the domain of homeless people and nobbers. For the time being, I was both. With no cash for anything other than a bottle of water and some cheap dodgy street food, the night fell and still I wandered the busy streets. My feet were aching, my shoulder straps on my daysack cut into me - it was like being back at work. I worked out I must have covered about 20 miles just constantly drifting aimlessly around the resort. Like a prison escape I sought familiarity, so returned to the hotel where I'd previously stayed. I knew they left beach loungers out overnight, but also knew they had security patrolling the beach. Even so, my options were limited to dragging a lounger into a patch of dead ground just off the beach entrance to the hotel, and settled down for the night with my daysack as a pillow and my damp towel as a sheet.

I should have reported security. No-one caught me. I woke chilled to the sound of soft surf. The yellow hued morning sun pierced my eyelids. I opened them to be greeted by a million dollar view of the white froth of the Caribbean Sea playing 'fetch' with a stick that it dragged up and down the soft white beach. The simplicity of living like this warmed my soul, but reality hit when my stomach gurgled its disapproval of not having anything to digest. Starving, with only another fourteen hour bus ride to look forward to, I walked to the bus station and caught the decrepit bus. Sleeping on a daytime trip was harder and the people on board far busier. With my Spanish limited to 'Dua cerveza por favore' conversation with the locals was hardly inspiring, but as I've often found, strangers of any culture are usually welcoming and with smiling a universal language, I was offered fresh fruit and bread to postpone starvation.

Arriving in Belize City at night is fraught with danger. Street lighting is poor to say the least, and as in many major cities, nighttime draws out those with less than honourable intent. I was in a dilemma. I was stuck in the city with no money to get back to Holdfast Camp 73 miles away. I didn't even have enough to get me to Airport Camp, which I'd passed about half an hour earlier. I'd have alighted there if I hadn't been asleep.

The only transportation option was a taxi. But without money there was little hope of getting a ride home. A taxi driver did approach me and I told him my destination. It was perfect for him. He lived in the capital Belmopan, so it wouldn't be too far out of his way. Rather than going home with an empty cab he would gladly take me. The problem was my cash situation. With resourcefulness oozing out of me like dirty sweat, I offered him all I had - a bag of clothes and a bottle of duty free tequila. He was more than happy. There is a lot to be said for a barter economy, although I did start to wonder whether I was to end up in a bush with a bullet in the back of my skull as on the journey he produced a pistol from his centre console. He started waving it at me.

When a man waves a gun at you, there are usually three reasons for doing so:

> 1. He is going to rob you. With my pathetic cargo that I'd already promised him gratis, I could only come to the other conclusions.
>
> 2. He is going to rape you. With blatant disregard to the 'No Entry' tattoo on my virgin arse.
>
> 3. He is going to kill you. I admit, this did play on my mind somewhat.

Thankfully, my worry was unfounded. He told me it was to repel any armed bandits that may try to hijack us as was common on this road at night.

Oh OK then. That stops me worrying.

***

Getting back into the swing of work was difficult. The Troop Officer hadn't been impressed with the stunt I'd pulled, although I think he was more pissed off with the BSM for overruling him. In an act of revenge, he took my name off the list to go the St George's Caye for a week's diving. Instead, he put me on a week's guard duty. I continued to thrill seek, and it became clear that my over-indulgent lifestyle was affecting my fitness. Running became harder after a night drinking from the fountain of foolishness. So like any good bootneck I gave it up, running that is. No longer was I one of the stronger runners on the morning runs, I was now being overtaken by those who once had been trailing in my wake.

I was struggling to balance work and pleasure but with only a month to go I was on my run down period (RDP) I wanted to squeeze every last ounce of pleasure from this country I'd fallen in love with.

With the Belize tour coming to an end, our final evenings were spent reminiscing on our time together. It concluded with a dinner, and smart casual clothing was the order of the day. I couldn't think of anything smarter or more casual than my kiddie-sized plastic Teenage Mutant Ninja Turtle outfit that I'd purchased in a drunken shopping spree when in San Francisco.

Wrapped in green plastic, sat in a restaurant exuding Central American humidity, I sweated 'til my crotch started to rot. If I was a food critic for Ninja turtles, I wouldn't advise them to eat here.

If we think of stupid ideas, we may think of entering a bear cage smeared in Salmon paste or giving a pet dog passion fruit*. What is even more insane was to dip my nob in the hottest Belizean chillie sauce known to humanity, just to prove how tough my nob was, a key attribute, apparently, for a Ninja Turtle. Being a complete nob is one thing. Having a complete nob is another, as the top layers of skin on my glans peeled away, burnt from the ferocity of the sauce. Pain dealt me a swift and mighty blow. The only pain I'd ever felt of such magnitude was getting my raw blisters sprayed with iodine in training. However, this was on my nob, and self inflicted. After picking himself off the floor from laughter, my trustworthy friend Pete passed me over a cup of water to soothe and cool my shrivelled chillie gherkin. It would have done the trick, if it was indeed water. But it wasn't. The sadistic untrustworthy bastard handed me a cup of '151' proof white rum that he'd poured from an adjacent bottle. So now my near-raw nob was dipped in near-raw alcohol. The pain intensity went up another notch, and as tears poured down my face I ran stumbling to the restaurant toilets. On my tip toes I stretched my burning todger over the edge of the dirty cracked sink and washed it under the cold tap like a good first aider allowing the water to cool my burn for twenty minutes. Well it was more like five. I got bored, and as it had reduced to a manageable level of excruciating pain rejoined the lads who I have to admit were pretty accurate in their description of me being a complete bell end that, under the circumstances, wasn't particularly accurate.

Although full of fun, laughter and merriment, an end of tour dinner/piss up is often tinged with sadness, the end of another chapter in our lives. The Belize dinner would definitely be so for me. It would probably be the last time I would ever see the majority of the 7 Battery lads, which was a tremendous pity. These lads, that I'd come to know so well over the previous six months, upheld the commando tradition in every sense. To a man, they were as professional as any bootneck I knew, and wore the green beret with equal, if not more pride than the average Royal Marine. Once passed out as a Royal Marine, those who chose could slacken off, comfortable in the knowledge that they had a job as a commando for as long as they wanted. To get kicked out of the Corps and lose the right to wear the green beret, murder most foul would have to be committed, or you were the ultimate biff, forever being charged and one who should have never passed out of training. The army commandos, on the other hand, always had to maintain their fitness and soldiering skills to the highest standards. Those that didn't would be sent back to a 'crap hat' unit. They would always wear the insignia of a commando, 'the bread knife' but they would lose their beret, instead wearing the colour of their new host unit, and there was no worse punishment than going from commando green to navy blue headwear. I was so impressed with my time with 7 Bty 29 Commando Royal Artillery that I seriously considered transferring over, yet the Royal Marines had other plans for my future.

*Believe me, this is a stupid idea. It is the basis for that famous philosophical quote: 'Allow your dog to feed upon passion fruit and you will be late for work in the morning'. I once noted my dog eating passion fruit that had fallen from our tree. I should have stopped him sooner but he seemed to enjoy them. The following morning, I opened the door to the laundry where he slept, to let him out for his morning ritual. It was a bit late. The poor thing had shit himself inside out and had done a dirty protest all over the tiled floor and walls. Washing a dog matted in shit and passion fruit seeds does not make a good start to the morning.

'At one time leadership meant muscles;
but today it means getting along with people.'

~ *Mahatma Gandhi, 'Father of the Indian nation'*

WITH SIX WEEKS POST-BELIZE leave, I wanted to jet off somewhere equally fascinating, but with an imminent personal financial crisis on the horizon, austerity measures forced me into spending my leave in the exotic surrounds of Norton Manor Camp. My colleagues in AE Troop were still working, yet as a free agent I spent every night ashore with Pete.

From initially being a climbing dilettante, obsession would take over once the stars came out. I'd managed to night climb my way around the crumbling buildings of the locale à la Freddie Spencer Chapman - if such a British hero as Freddie did it, I thought it less foolish - and there wasn't a first floor window of Taunton town centre I'd not clung to, nor the 20m high wall en route to Kingston's nightclub where I taught younger lads to jump 10ft across to the top of a barbed wire covered lamppost in order to garner attention from young ladies, yet only succeeded in attracting the attention of the police. Unfortunately, one of my protégés, Bez took his climbing skills north and scaled the side of a large manufacturing unit in Newcastle quite successfully, until a ledge gave way around 15m up. It wouldn't have been as funny puncturing his lung, breaking his ribs and jaw and having wooden teeth inserted by a maxillofacial surgeon if it hadn't happened two weeks before his wedding. When a groom needs sign language for his reception speech and his wedding cake has to be liquidised it is quite a memorable wedding and for fellow bootnecks, quite hilarious. These impromptu climbing expeditions were the drugs I needed to alleviate the monotony of camp life and all I had to look forward to between the end of the Belize tour and Christmas leave, where I would, once again, indulge in a skiing holiday, this time in Italy. Oh, and the small matter of preparation for the next Junior Command Course.

The Junior Commando Course (JCC) is the eight week leadership programme for promotion to corporal. Somehow I'd been deemed suitable for this. Usually for lance corporals, the course can be given to marines, as the rank of lance corporal in the Corps is an honorary one, unlike the Army, where you have to pass a course to get your first stripe.

I'd never really sought promotion. Indeed, I was only a candidate for promotion because Scouse, my Troop Sergeant in Northern Ireland, wrote out my application and told me to sign it. He'd seen something he liked, and so evidently did others. Despite my best intentions, I kept getting good grades on my six monthly reports and a distinguished mark on my Assault Engineer course had catapulted me up the JCC loading list.

I had to seriously consider whether I was a worthy candidate. I could hardly look after myself let alone a section of others. I'd just turned 21 and usually only those in their mid twenties or even early thirties with years of experience were loaded onto the JCC. I'd spent the last year trying to kill myself through booze and women and now I was expected to be part of the leadership strata.

What example could I set, when I couldn't even go on leave without locking my locker?

How could others look up to me when I would start a fight, absolutely ming-monged from flaming

Sambucas, against another bootneck who'd shouted abuse at my ex girlfriend?

Probably the fact I'd head butted him into a shop window, then dazed and confused woke up with him straddled on my chest punching the fuck out of me, only to be told by my mate Jammy that, in the best tradition of slapstick comedy, I'd walked backwards across a road, only to trip over a traffic island, fall on the floor giving my victim the opportunity to come over and exact swift revenge; should tell me I wasn't fit to lead.

Maybe waking up late for work the next morning with a head like a bruised melon, and in a panicked state quickly changing into my rig, jumping on my bike and in the early morning frost, peddling as quickly as I could leaned my bike right onto the parade square, only for my wheels to lose their grip causing me to slide into the first three columns of Alpha Company who were paraded outside their company lines being given a bollocking by their sergeant major; indicated I wasn't yet mature enough to climb the ladder of rank.

The fact his bollocking turned to me as I picked my bike up from the pile of felled bodies and then cut short when he saw my melon head, so ordered me to go sick; should tell me to decline the invitation to seek promotion.

Dinger was also on the course, and with others I discussed my concerns. They accepted me as a 'good time Charlie' but they advised that if I turned down this opportunity, it may be a while before a new one came along. The Corps didn't take kindly to their decisions being questioned, and as it was classed as a promotion opportunity, should I abstain I would go to the bottom of the loading list and God knows how long it would be before my name, forever ringed in red pen, surfaced again. With this in mind, I duly accepted the draft, and with CTCRM my destination straight after leave I thought it best prepare properly and go for a week's skiing holiday in Italy.

Sauze d'Oulx, known as 'Lousy Souzy' within the skiing fraternity, is like Magaluf on snow, and our après ski continued for so long it became avant ski. Dinger, Frankie the ski guru, and Pete came along and together we ensured that when we weren't on the slopes egging down every dangerous slope as quickly as the drag on our fancy dress would allow, we'd be clubbing and embarrassing ourselves in the many bars around the small village attempting to encourage any female to engage in poor quality sex (I am quite rubbish at it - ask anyone).

Upon the holiday ending, I should've really knuckled down to get some study in for the JCC. Dinger had taken this option but Frankie's parents lived in Brussels and as he was spending the New Year there I thought his open invitation would be a good way to see in 1990. With Belgian beer being so strong and relatively cheap there was really no point in studying for a career promotion.

La Grand Place in Brussels is the equivalent of Trafalgar Square where, on New Year's Eve, it becomes a swarming hive overflowing with throngs of nip-nosed revelers. The beautiful baroque and gothic buildings that walled the square were decked out in traditional seasonal decorations with a huge Nordic pine standing dominantly classical over the plaza. Acrobats gave demonstrations to the happy crowds who were bamboozled by the idiotic bootneck who joined them to do rubbish cartwheels, forward rolls and a handstand that would embarrass a four year old. The local gendarmerie were quite casual in their dragging away of this fool, and with a warning not to do it again, they let me go to enjoy more sickly sweet ales.

Having done no preparatory work whatsoever, I started the pre JCC course way behind the eight ball. The two further weeks during this phase would hope-fully get me to a level near where I should be, and I found the corporal taking us very astute, incredibly intuitive and gave us the inside hints on how best to pass the course.

I was the youngest on the course and Lord, how I felt it. Feeling like a nod again in the company of men who I immediately considered my superior, I felt a little embarrassed to be wearing the single stripe of a lance corporal and the number brassard to indicate I was on the JCC.

I never thought that after basic training I'd have to complete the Endurance Course ever again, yet the JCC week 1 culminated in rediscovering its pleasure. This time we ran as a section, which wouldn't have been so bad if we hadn't had to lug a telegraph pole along with us. Even so, I found it far easier than when a nubile recruit. My body was no longer the train wreck from incessant beastings but now strengthened by age, passive smoking, alcohol and the self confidence of just being a bootneck, I actually enjoyed wading through Peter's Pool on a cold January morning, looking like I'd been dredged from the bottom of the Manchester Ship Canal.

Theory I could handle all day long. The classroom tests we had, I passed comfortably. It was out in the field where it counted that I felt I was lagging behind. I'd rarely given patrol orders in a unit and only then the quick version. Here, to fully prepare, write and deliver them was a whole new ball game. The JCC accommo-dation was always the last to extinguish lights and plenty of students would burn the midnight oil in the lamp of diligence. I, on the other hand, figured that the crux for me was passing the fieldwork tests, something that couldn't be revised for; therefore enjoyment became my nightcap, spending most evenings in Exmouth's 'Sam's' nightclub trying to dance myself into a promotion.

I knew we wouldn't be treated particularly well. CTC has a knack of turning decent people into wankers and the DS of the JCC training team ensured we had an uncomfortable course. I thought running around Woodbury with a rifle above my head was in the past. I was wrong. I thought having to wait all night for a vacant washing machine was part of the routine only nods did. I was wrong. Indeed, we were getting a worse deal than the nods. Dartmoor had been closed off to recruits due to the weather conditions, yet we continued on to do the 30km night navigation exercise in visibility similar to walking round in a blindfold. We were commandos, recruits weren't; no matter what the conditions, we'd crack on.

Patrol exercise started early Monday and finished at first light on the Friday morning. In between we prepared, listened to, and gave orders before executing the patrol as required. There would be no point in packing a sleeping bag as we would not feel its amniotic comfort - we didn't sleep. People never believe me when I say this, but I can categorically say we never purposely slept. We may have nodded off, but immediately our survival instincts kicked into gear and we awoke with a start. We dare not be found sleeping, that would be a warning. Two warnings and we would fail the exercise. It was the hardest thing I'd ever done. What we had gone through in training was easy in comparison. What I'd so far encountered in a commando unit paled into insignificance. We were being trained to be leaders and pushed beyond our limits once again.

My throat is so sore I find it hard to swallow. I've lost my voice. When my command appointment came, my orders were given in a hoarse whisper far too quiet for the guys who sat under a bivvy listening, instead, to the hard drumming of rain on the sheet above their heads.

I am again trying desperately to stay awake for the orders that Kev is now giving. Despite the let up in the rain, I'm wet through. The coffee I've just drunk warms me for only so long but the damp seeps through to gnaw at my aching bones. It could be worse; I could be wearing a helmet. It's 0430hrs. I haven't slept now for nearly 67 hours. I thought it impossible to stay awake this long, but I've literally not stopped since we jumped out of the back of the four tonne trucks at four firs car park. I nearly fell asleep while I helped Kev build his model for these ambush orders, the longest set of orders, the hardest time to stay awake. I felt bad when he checked my work and noticed I'd inaccurately placed the FRV. I looked at the map again through blurred eyes and the cone of light from my head torch to see that he was right. Of course he was right, he was a switched on cookie yet his rebuke was received with the same humour as it was given. As I sit here, I wiggle my toes to both stay awake and to get the blood circulating again. I stare hard to get air into my dozy eyes wishing I had those legendary matchsticks to keep them open, and pick my empty nose just to keep my arms moving. I look around the shadows. Marv is staring vacantly towards the model that is visible under the portable floodlights of head torches, his camouflaged face covering his wan pallor. Deano looks the same. He loosely holds a pen but his writing is only scribble, trying his best to control his fingers that just want to sleep like the rest of him. I wonder whether Kev's words are just bypassing their brains, as they are mine. In the darkness the DS watches. Although virtually invisible, we know he is watching us as much as listening to Kev who points his stick to the objective that I struggle to focus on. As it's the umpteenth patrol we have been on today we pretty much know the score but we have to go through the whole process, fully, each time. No quick battle orders here.

'Right, any questions?' We know this mantra means that the orders are finally coming to an end. 'No? Well I've got some for you.' Again, Kev is going through the motions to keep us on our toes although the questions are preloaded.

A shuffle sounds as we try to sit upright to keep ourselves alert.

'What are the actions on at the FRV?' Kev pauses, then with a four fingered point he nominates towards the darkness, keen not to have his head torch waving a beacon to any supposed enemy. 'You at the back.'

'Me?'

'Yes you numpty, who else?'

'Well seeing as though I'm the DS probably another student would be a better option.'

Such mistakes give us the jovial shot in the arm we need.

Finally, after the questioning ends, we can stand from our sitting positions. With the task at hand understood, we now add that extra layer of clothing - humour, by ribbing each other about our battle against sleep and now labeling Kev as a sniper. This banter radiates that extra warmth - camaraderie. Morale again raised, we carry out our pre-patrol checks before we head out to lie in a tree line for the next 3 hours where we will fight sleep then the enemy who will likely show just after first light. As long as it stays dry it should be bearable. I should have kept my thoughts to myself. The first drops splash the leaves in front. You're not a proper bootneck until you've had trench foot anyway…

I failed patrol exercise not for any other reason than I failed my orders. I'd passed the execution of the orders, which was a miracle considering no-one heard what I'd said. My laryngitis had come at a most inopportune moment, yet the serrated whispering with a box cheese grater stuck in my throat held no sway with the DS. He failed me, reasoning I didn't convey my orders clearly enough. I considered giving my next set through the medium of mime.

Weekends on the course were always a chance to catch up with reality; reality being sleep and enjoying the 'Turk's Head' pub amongst the impressionable nods. Even a trip to 'Tens' nightclub brought back memories and with Reni and Reedy I was invited back to a nurse's party held in some unknown location at some address, somewhere within the vicinity of Exeter.

Reni and Reedy were truly thick as thieves. Both outrageous characters of 40 Commando's Alpha Company, they shared many incidents on the wrong side of military discipline. Reedy was a man to whom tasks were completed easily, which emboldened him to a swagger of personable arrogance. He didn't take orders well, which continued on the JCC, evidently not the best way to get a gold star. Reni was a legend. Although the only cool bloke I've ever known called Tarquin, he was incredibly fit, even for a bootneck, but one who partied harder than Peter Stringfellow.

In one incident, Reni was in the Cornish military camp of Penhale preparing to go on a reconnaissance concentration exercise. The next morning would see him and his fellow recce troopers yomp twenty odd miles over the Cornish countryside to an objective, so a good night's rest would be an ideal way to prepare. But sleep would be better should the lads tire themselves through partying with the surfing fraternity in nearby Newquay. Reni had ended the night alone and rather inebriated, therefore started walking the seven miles back to Penhale camp attempting to hitch a lift struggling to both raise a thumb and prevent his bag of chips from falling. Luckily, a lorry driver saw him and picked him up. Getting into the warm cab, Reni fell into a drunken stupor. Awoken by the lorry driver, he alighted the cab a little disorientated. The reason for his confusion was simple. The lorry driver had misheard Reni and dropped him in Penzance not Penhale. No worries, Reni could run it back while attempting to hitch a lift en route. Off he ran, in his shoes, jeans and flowery shirt. He ran, and ran, and ran some more. He eventually arrived in Penhale twenty eight miles later, just as the lads were getting up for breakfast to get the energy for the twenty mile yomp that would commence in an hour.

Back on the JCC, the nurse's party was going well until the cake was brought out. Candles were blown and girls clapped politely. Trying my best to be polite in such company was proving difficult when so hungry, and as the cake looked terrifically tasty, survival instinct took primordial vengeance upon the victoria sandwich, excavating a lovely chunk into my gob. Reni yanked my hand inadvertently wiping the majority of the cream and jam around my face. In return, I threw a slop of raspberry jelly, smudging it all over his wiry hair. Reedy probably didn't expect the load he received either, as he returned from the toilet. Smearing ourselves with expensive cake and participating in a micro food fight open to only three participants, wasn't the best way to endear ourselves with the party host who, to put it mildly, went rather nuclear, throwing us out in a giggling heap of crumbs and butterscotch cream.

***

It was early February 1991 and 'Operation Granby', the British codename given to the 'First' Gulf War, was in its offensive phase.

We British are extremely shit at naming operations. It seems we conciliate our acts of violence by giving them the most foppish name possible. Admittedly the gung-ho names that the United States dish out like Operations 'Desert Storm', 'Urgent Fury' and 'Kill a Terrorist' are a bit over the top, but 'Op Granby' was named after someone of the aristocracy, who was commissioned as a colonel, later to be a lieutenant general in the Seven Years' War. Worse still, in the not too distant future I would be deployed onto 'Op Warden' named after someone who sticks penalty fines to your car windscreen.

Although nearly 55,000 British military personnel were deployed to the Gulf region, only a few were bootnecks. It was a huge disappointment throughout the Corps that we'd been overlooked for the conflict. We had a few detachments dotted about - M Company from 42 Commando was entrenched in a swanky hotel in Dubai. Seemingly their operational mission was to strengthen international relations by shagging as many foreign airhostesses as possible, a task, by all accounts, they were hugely successful in undertaking. A few lads made it into the national media by fast roping into the UK embassy in Kuwait, but in the main it was a mechanised war, tank battles and mechanised infantry dominated the barren desert landscapes, while in the air the coalition destroyed anything enemy-like with embarrassing ease. The lads from the RHF, who I'd not been impressed with a year before in Belize, won battle honours through their actions. It taught me a valuable lesson in not judging through arrogance.

There had been a few buzzes knocking about that 3 Commando Brigade was likely to be deployed, and we lived in hope that the call up would come. On the Monday morning we thought it had. With the morning's drill postponed until an important clear lower deck had finished, we marched to the drill shed expectant on having the course disbanded, and all being returned to our units to go and win the war for Britain and be home in time for tea and medals.

The course officer looked stern. Obviously important, we listened with anticipation to the odd introduction to his talk. It wasn't about the Gulf per se but about leadership and setting examples. The talk was going way off course to what I was expecting. It then hit me as he talked about our responsibilities, both in and out of uniform.

'Over the weekend,' he continued; 'three men, now known to be on the JCC, destroyed a nurse's party, vandalising the bathroom.'

My heart pumped from my chest. Vandalism? If throwing a cake around was vandalism I was guilty, but smashing up the bathroom?

The course officer continued, 'It won't be difficult to identify the miscreants and it should be noted by those responsible that we will look favourably should they come forward to admit their offences.'

It was clear they would find out. We'd been putting nurses in headlocks for photo opportunities.

The three of us reported to the camp RSM. He was a fearsome man, with a fearsome reputation, and fearsome ears sculpted from far too many fearsome rugby scrums. We admitted to throwing cake but could not assist with the bathroom vandalism enquiries. Without really considering our plea, we were given the option of being kicked off the course, and possibly putting any future promotions in jeopardy, or paying £300 to make good the damage. Angry at having to fork out such a sum, I tried to recall what expensive items she had in the bathroom. Possibly the doll shaped toilet brush cover was an heirloom from a mad aunt with some mental disorder? Or the damp infested shower curtain had some value to the communicable disease research institute? We'd been mugged. In revenge for our cake destruction she'd got one over on us big time, and we deserved it. With a disciplinary RTU hanging over our heads should we transgress just once more, we continued onwards, with empty wallets, trying to avoid nurses and delicious cake.

The course continued unabated. Some students just cruised, doing as little as possible to scrape by. In the opposite corner, those who were keen as mustard, intent on claiming the best mark possible. Most were team players, akin to inquisitive puppies keen to get involved in everything, then unfortunately the few students who had the duplicitous deceit of a parsnip masquerading as a roast potato - surprising, yet wholly distasteful. They would often step back from assisting anyone until the training team showed up, then they would be become Mother Teresa in DPM, raising their hand to any request, volunteering to carry the heavy loads or assisting with stores, with little truth being played out in front of the instructors' note-taking gazes. Unfortunately, some of these sailed through the weeks with little trouble, destroying the very values they had been taught by the Corps, so I eventually viewed the whole event as a bit of a sham, reverting to play the simple 'course man'.

As the weeks progressed, we could see our progression; each phase of the course carefully mapped out to allow us to self-analyse our journey towards the culmination of the course - the final exercise in Sennybridge, Wales. As we left Lympstone en route to the misery of the Welsh gloom we knew a good night's sleep was now a distant memory, something we couldn't even dream about for the next week so, as per corps tradition got our nappers down on the bus, proving not only that bootnecks are masters in the art of sleeping at the drop of a hat, but also that 'convoy cock' - the gaining of an erection through the vibration of transport - was not a fable.

With more patrols, an uneventful, insomnious, yet energy sapping defence phase, and another liberal smattering of dismal weather; the final exercise ended with a speedmarch back to Sennybridge camp, straight after a dawn attack. I was carrying the spare ammo for the troop and had lost count of how much I had in my daysack. The extreme weight in my daypack told me it was a lot. I hoped I'd get rid of it all once the attack was over. Wishful thinking that never materialises always brings on a slight case of dismay. I asked one of my section oppos, one of those sneaky fuckers who played up for the DS, to take a few boxes.

'Fuck off, ' was his tacit reply, one that told me that I wasn't going to lighten my load any time soon.

We were rushed into the speedmarch with no time to even take off the warm clothing I'd worn to protect from laying up overnight. If there is a way to make a speedmarch difficult, then I'd done it with an A grade pass. With far too many clothes and with more ammunition than Israel, I was already 50 metres behind my section so had to sprint up a mud bank to the group before we'd even got into rhythm. My mind pulsed back to training and the difficulty I had back then. Even being fitter and stronger, my legs hadn't lengthened by any considerable amount and I have yet to see any Olympic middle distance runner choose a heavy pusser's helmet as headwear of choice - their bullet-proof capabilities ideal for boiling one's head, so collapsed exhausted at the finish. Many had dropped

behind and for the first time ever, I'd seen Reni knackered.

'What the fuck's going on, Corporal Time? I
thought of you as one of the fittest?' asked my DS, his
own body steaming under the frost of early morning.

I didn't even have the energy to lift my daysack after I'd thrown it on the floor, 'Lift that bastard.'

He picked it up. 'Why didn't you share it out?'

What could I say, stitch up the jack twat? 'Didn't
have time, Sergeant.'

'Well good effort on phys, shit effort on admin,' he
said, walking away.

I didn't need a 'good effort'; I needed my head examined for carrying that lot in the first place.

Although I'd passed the criteria tests there was no
guarantee I would pass the course. I wasn't a strong
candidate and my misdemeanours earlier in the course
meant I was frequently under the spotlight. But my ill
confidence was unwarranted. I managed to pass. At the
tender age of 21 years old, I became the youngest
corporal serving in the Royal Marines.

***

The call eventually came. 3 Commando Brigade would
be deployed to the Middle East. A bit late, Op Granby
had finished. The Iraqi Army was licking its wounds
under the pretense of Saddam's propaganda machine
claiming it victorious. Despite him being given a taste
of coalition hospitality, he regrouped then continued his
savagery, turning his attention to the minority ethnic
groups around the country.

Instigated and encouraged with a call to arms by
the then US President Bush, disaffected ethnic groups
under the notion that the US would support them,
attempted revolt. Realising the revolts weren't quite

what the US administration had in mind, and seemingly not in the best interests of the US, the Americans withdrew, leaving the belligerents to fight the Iraqi Army with no support. In the south, Saddam not only took violent reprisals, but also drained the marshes to deprive the Madan marsh Arabs of food and refuge in response to their support of the Shia uprising in Basra. In the north, he turned his attention to the Kurds. Swift retribution was served. Saddam's beaten army was still mightily equipped when compared to the Kurdish militias. While guerilla tactics could accommodate small scale missions, Saddam's army concentrated on destroying the Kurdish people. Whilst the UN had administered a no fly resolution, Iraqi air arms strafed the Kurdish people from above. Armoured divisions rolled north undeterred, intent on dishing out a great big dollop of genocide, bulldozing Kurds into the inhospitable winter mountains. It was here we were to be sent on Op Haven - immediately renamed Op Shaven Haven.

'Only the dead have seen the end of war.'

~ Plato, Greek philosopher

JOINING 5 TROOP was a less than auspicious return to Bravo Company. I paraded at 0815 as I always did in Support Company. Bravo Company paraded at 0800.

I hadn't even officially joined the troop. Prior to leave I'd asked to return to a fighting company as I wanted to 'learn my trade'. AE Troop didn't have the capacity to retain so many corporals; therefore, my request was readily accepted. Yet even I didn't expect to deploy on operations without even being formally introduced to my new colleagues.

As I stood in the 4th 'supernumerary rank' with the other JNCOs, I was pleased to see Deano, a guy who I'd become close friends with on the JCC. Looking around the troop I saw him as the only familiar face. I looked across the company parade and saw old mates - Simmo and Carl - still there clinging onto life, stalwarts from my previous time in 'Beef Bravo'. But here in 5 Troop, just about to go on my first operational tour as a newly promoted corporal, I didn't even know who was in my section. So by way of formal introduction, before boarding the coaches to transport us to the air head for the RAF flight to south eastern Turkey, I called in my section for a quick 'hello' with a promise that I'd talk individually to everyone on the coach.

Personally, I think it vitally important as a leader, or a manager, to know what makes people tick. I'm passionate about sharing their backgrounds, ambitions and to allow them to express themselves. Certain military doctrine wouldn't necessarily support such individualistic methods - guys are there to take orders, but a great leader is someone who can get people to do things they don't necessarily want to do, so my interpretation was to find what made a person behave a certain way so I could play to their motivations and gain better results. A ten-year marine would hardly have the same ambitions as a raw 18 year old straight from Lympstone, so varying my methods to suit was needed to attain the necessary result. As it transpired, apart from two guys straight out of training, I was the youngest in the section. With the rest all drafted in from various parts of the Corps on that day, it was my job to form a strong amalgam from hitherto strangers - and quickly. We'd be landing in an operational area in about six hours, and would not know when, or if, we would return.

I was keen for my section to be cheerful. The Royal Marines promote to every recruit the four 'qualities of a commando' - one of them being cheerfulness in the face of adversity. I tried to instill this even in the more comfortable times. Someone who works with a smile on his face is far easier to handle. He is more confident, will have intrinsic motivation and will more likely to assist than hinder. This communal cheerfulness would assist in bringing my guys closer as a section. It would have been preferable to see more of them in this fledgling relationship but now, as part of the leadership group; I had to attend briefings for our arrival in theatre.

Prior to leave, while in AE Troop, we had packed as ordered. We were going into the mountainous region separating Iraq from Turkey, and therefore we would be kitted up as if we were going to a cold weather environment. 40 Commando had just completed their first Norway deployment so this would present no problem. Therefore, it was a little surprising when we debussed, like kids on a school outing, on the southern Turkish border into an area as flat as Norfolk where the temperature was 30°C+. It did seem we had somewhat over packed for the occasion.

This was the first time I'd been on a deployment where no end date was set. Northern Ireland tours, exercises and training activities all were programmed in with a preordained return date. Not knowing the duration of a tour is a strange feeling. For those with a family, it must be mental agony not knowing when they would see their loved ones again, no calendar to mark off 'days to do', no 'see you soon' messages, just a timeless separation in a doctor's waiting room where we didn't know whether we would ever be seen. Moreover, the CSM instructed us all to write 'last letters' - heart wrenching messages to be passed onto our loved ones should the worst happen. I sat in front of my 'bluey' - the Forces blue writing envelope that presented licking glue so acrid it could only be made of boiled down horses fed solely on lemons - with the intention of creating my most important piece of writing, yet I just stared blankly at it as if on the calculus section of my maths 'O' level paper. I couldn't even begin to understand the feelings of a father writing to his wife and kids whom he may never see again. I didn't know what was sadder, their feelings of unconditional love presented awkwardly as words or realising that I had no one to write to.

Intelligence reports coming from the mountains were grim. The reconnaissance parties, already positioned, reported of the humanitarian catastrophe that awaited us. The Kurds, now classed as Internally Displaced Persons (IDPs), had concentrated themselves into large camp areas where UN and military aid flights dropped life support. In our area of operations, the

Çukurca camp held 400,000 Kurds suffering from the conditions. Typhoid, cholera and dysentery were rife and up to 2000 people a day were dying. Those trying to escape into Turkey were brutally beaten back. Turkey had its own problems with Turkish Kurds so were hardly accommodating. The IDPs were trapped, corralled between an unfriendly border and the Iraqi Army. They required humanitarian assistance most urgently. It came in the form of 3 Commando Brigade Royal Marines, part of the Coalition Task Force.

We'd been trained to fight. We hadn't been trained for this. We would have to be flexible, resourceful, compassionate yet robust, and be vigilant for any situation that arose. We were to be combat social workers - a weapon in one hand a handkerchief in the other, an iron fist in a velvet glove. Even without any formal training, surely we were the best candidates for the job.

***

It was time to deploy into Iraq. The air tasted of AVCAT fuel. The cerulean sky black with the huge flock of rotary winged birds, their blades chopping the air into a cacophony of Vietnam war movies - the familiar sound of US UH1 Iroquois 'Huey' helicopters buzzing around the large behemoth CH53E transport helicopters. En mass, a fleet of USMC CH47 Sea Knights landed to transport the rows and rows of waiting Royal Marines kneeling behind bergans that were bigger than the wearer. It was an impressive sight and as the first laden helicopters dissolved into the mysterious mountains of Iraq to our south, I could not help but feel a little apprehension, yet excited and immensely proud.

Darkness was falling. We had waited patiently on the landing sites (LS) for the other companies to deploy, the helicopters working tirelessly to ferry troops over the border. Bravo Company was the last to insert to their respective location. Most of the company had already left, but with our group being part of the last two chalks, news filtered through - a gen buzz Royal - that we may not get over the border that evening but have to wait until first light the next day. The pilots were up to their maximum flying hours and refueling had seemingly not been made a priority. The swear box would have been full within five minutes should we have made one.

A Sea Knight landed adjacent to our LS and the Loadmaster gave a firm wait as we started to rise from our kneeling positions. From the cockpit climbed a pilot who beckoned our Troop Officer forward. Stunning in every way, the pilot's blonde hair immaculate, even after being crammed into a bespoke helmet for the past eight hours, she smiled a toothpaste advert smile to our group and like a pubescent lovesick schoolboy I wasn't the only one to think she had smiled personally to me. She gave a thumbs up to the Troop Officer. I've never seen so many men as eager to get into the back of a vehicle controlled by a female, leaving a snail trail of drool along the fertile plains of south east Turkey.

Once in the cab, I was handed a set of communication headphones.

'You guys realise we are running on fumes here?' said the Loadmaster, his New York accent as thick as the lamb hotpot I'd just consumed in the new 'boil in the bag' ration pack. 'We're only taking you 'cos the pilot is one crazy ass.'

The crazy ass goddess of a pilot laughed, tinny through the headphones.

'Yup, it'll be a miracle if we get back,' she said without a hint of humour. 'Geez, you guys are going in so deep that I wouldn't be able to sleep tonight knowing you had your brothers out there who need you to complete the mission. You're a risk worth taking.'

Despite having the Gorgonzola of all cheesy comments in her repertoire, I had to admit that if she happened to like Manchester rave music, shop brand bourbon biscuits and short, gobby northerners, she was the perfect woman.

We checked our maps in the aircraft, looked down to the mountains below and saw we were getting dropped at the correct grid.

One chalk from Alpha Company was also getting dropped 'deep', even deeper than us. They made their clandestine landing hidden by the darkness of mountains and blackened skies in true commando fashion. The chalk commander checked his map sheet to confirm their location. After a lengthy confirmation it was found they were at the right grid reference, but -

On the wrong map sheet.

In the wrong country.

To make this the mother of all fuck ups, that wrong country was Iran.

Border incursions are hugely sensitive issues at the best of times but a low flying radar-avoiding US helicopter full of British commandos landing illegally in Iran - from Iraq - would, I imagine, kick the recession back home from the front page. Once realised, the chalk was recovered and stealthily withdrew from whence they came awaiting the shit storm of ridicule from the other companies.

Back in Iraq, our location that was 'so deep' was in a valley system on the banks of the raw Great Zab River. Here the river had mountain power, huge bedrocks making her spew white froth, angry from the melting mountain snows. Evidence of Iraqi Army presence was everywhere. Old grenades lay rusting along with used ordnance, discarded respirator gas masks, and minefields, not only of real blow-your-legs-off mines, but even worse, oh-that-is-fucking-disgusting human shit.

As the darkening sky fell into bed with the blackness of the land, we set up a temporary night routine, sending out clearance and standing patrols to clear and dominate the area. In true colonial fashion we arose the next morning and paraded in the field. A first for us all, but the Company Commander reasoned an overt presence of discipline and strength would discourage an attack by any subversive elements in the area. It didn't, however, discourage our whinging.

The camp took shape. Renamed 'Rupert Brooke Camp' after the renowned war poet, we constructed physical defences, sentry posts and troop areas more akin to Rorke's Drift than a modern day Forward Operating Base (FOB). It was decided, as we were an overt camp, that stones would be laid around the area to make walkways and perimeters - all very 'Butlins'. To add a Hi-De-Hi element to this comedy, I made a signpost fashioned from a six foot iron piquet, with signs pointing to London, Baghdad, Bangkok, and one specially celebrating the Pakistani town of Kunt to point directly to the Troop Officer's tent.

We slept in our four-man arctic tents, perfect for the Iraqi sunshine, and ate rations designed for short term consumption. Any longer than a couple of weeks, supplementary fresh rations would be needed or constipation would rear its ugly (turtle's) head. Howeverer, there was a short supply of Tescos in the area and even the rear echelons were having difficulty getting hold of anything that would resemble real food. We could see the phrase 'I can't shit for shit' becoming rather popular for the next few months should we not even get a slice of bread.

Constipation was the last word on the lips of the other two sections of 5 Troop. Our early missions were to make contact with the local Kurdish elders, a hearts and minds exercise to enlighten and offer them confidence in our aim to assist their return home. We were their guardians with a promise that the Iraqi Army would no longer be a threat. I was a little pissed off at missing out on one particular meeting. Our section was the guard section so missed out on the opportunity afforded to the rest of the troop in making first contact with the Kurds. It was a blessing in disguise. Honoured by the meeting, the elders cooked a feast for the lads to tuck into. Traditional in its nature, the men sat crossed legged around a huge carpet on which lay plates of various dishes. Eating with their right hands, the lads partook in the dinner engaging through the interpreter on the issues at hand. Eighteen of them giving confidence to the Kurds. Eighteen of them that would return and within 12 hours be shitting through the eye of a needle from the food they'd consumed. Within 48 hours, a few of them had been taken to the Regimental Aid Post (RAP), the rear echelon medical facility, with agonising stomach problems and were losing fluids through either end at an alarming rate.

With Deano sat next to me on the 'thunderbox' field toilet, we tried to have a conversation with pants around respective ankles trying to envisage the beauty and majesty of the surroundings. It was an extremely difficult conversation to finish, as he kept interrupting my vision of these rugged mountains being the next dream skiing destination, by vomiting what can only be described as 'green shit'. As he puked, he convulsed as his bowels squirted what I could only imagine was more 'green shit'. It was desperate. Deano was as strong as an ox on steroids, and one of the fittest men I'd ever come across, yet here he was being slowly beaten to death by a microscopic bug. He was far from the worst. One of the Glaswegian RMR guys had fourteen intravenous drips squeezed into him for rehydration and was in a serious condition. This was a man who'd eaten fried mars bars and drank *Tennents* Lager for breakfast, but was now floored by a plate of goat. Tests returned from a specialist in tropical diseases in the UK confirmed some of the lads had dysentery, others a rare strain of cholera. My section had escaped the turmoil, so paraded around with our tops off, tanned and looking like proper bootnecks and not like the others - adverts for a new Band Aid record.

Bravo Company's mission was to secure the bridge spanning the Great Zab. The IDPs would use this route on their way home from the Çukurca camp once confident enough to do so. One troop would man the bridge checkpoint, one troop visiting the mountain IDP camp and the other would man the camp guard and do satellite patrols around the area.

I couldn't wait to get up to the IDP camp, and the 7km vertical yomp up there was exhilarating. The route was steep, along rough paths hewn from years of locals walking. Even for troops used to carrying heavy weight on routes that never seemed to be downhill, it was a workout. Our US Special Forces medic had collapsed exhausted on his first yomp with the Bravo boys. Comical as it was, he had to explain through extreme fatigue on how the lads were to set up the specialised equipment for his medical situation. Opening his bergan to extract the necessary IV set up, they found his bergan full of medical books. Stacks of use they are in a firefight. Subsequently on his next excursions to the camp, he travelled with a far smaller library on his back.

As we rose, the tell tale sign of human deforestation became more frequent. These particular trees weren't victim to the smooth clean lines of corporate loggers. Previously fully formed and healthy trees had been hacked to death by the displaced Kurds' rudimentary weapons, splintering them to give a grotesque and sinister entry to the camp's entrance. Strong acrid wood smoke with a tart hint of garbage strengthened as we neared an outcrop of rock that ricocheted the hubbub of noise away from our approach. Through a natural opening in a rock crag, the track took us to the open mountaintop.

I couldn't believe the four dimensional apocalypse that lay before me, one that shook from the soul the capital vice of comforted Britain - the slothful ignorance of things which matter - borne from a self-righteous entitlement to life. As far as the eye could see, amongst the craggy rocks, below the fug of wood smoke that hung heavily in the air, sat a human rubbish dump. Thousands upon thousands of makeshift shelters, made from whatever materials one could muster, carpeted the rise of the rolling mountaintop. Any available ground was strewn with putrid rubbish left by the huge throngs of adults that, as we slowly passed through, returned our comforting looks with vacuous, soulless expressions. It was a human catastrophe that could easily be met with casual ambivalence or even resistance to a slipper-wearing audience sat watching from their leather sofas waiting impatiently for the sports report. For us with every sense sledge-hammered by our unfiltered surroundings, it was a haunting scene, one that caused a feeling of anger, yet humility, that we were privy to scenes no human should ever witness, let alone be victim to.

As we were headed up to meet with 40 Commando Reconnaissance Troop and the detachment of US Special Forces, we couldn't take too much time with the IDPs. Yet as we walked further into the devastation, we couldn't ignore the hordes of approaching children, who shouted, smiled, and held their hands up wanting anything we could spare. Children with the least often give only what they can - a smile, so we flung the boiled sweets from our ration packs into the following mini crowd, berating the bullies as we did so. True to form it was always the bigger boys, so gave special treatment to the quiet ones at the rear without the confidence to ask themselves, and in their culture it was always the girls. The heart was twisted, not knowing what future lay ahead for these timid snotty nosed little girls no more than six or seven years old that stood immobile careful-ly watching us.

I was glad I remained a young, unmarried, childless adult. Should I have had my own kids, it would have been even more heartbreaking to see these children with no common language other than their pleading, sorrowful eyes.

The Non Government Organisations (NGOs) such as 'Médicins Sans Frontieres' were already set up with medical centres overwhelmed with illness. Recently created water points gave the IDPs clean water, and lines upon lines of women and daughters queued with buckets and jerry cans to receive their ration from the huge bladders that dotted the area. The smells continued to choke our nostrils, of wood smoke, cooking, wood smoke, rubbish, wood smoke, death. Even if we have never smelled it before, our primordial senses take over. We instantly know what death smells like, as we do the smell of burning, it is a survival instinct that stimulates our fight or flight response and as we passed a large white tent, death threw out its spine tingling aroma. Temporary mortuaries were set up and as we walked past one we moved out of the crowded path of an approaching stream of stretchers, each with a body shape hidden under a white sheet carried upon the shoulders of weeping men.

Arriving at the Recce Troop location, I bumped into a far more organised Dolly, who was now part of the intelligence network He was attached to the secretive Kurdish Liaison Team, commonly known as 'KLIT', aptly named as no-one ever knew where they were. It was also wonderful to see Nick, my best mate from the old days of 4 Troop. He was now a lance corporal in Recce Troop positioned permanently at the huge camp, close to the actual border with Turkey, manned by Turkish Army personnel, evidently conscripts from their ill fitting uniforms and confused looks. We liaised with the recce lads and the US Special Forces on issues surrounding the return of the refugees. The US SF guys we spoke to commented that some of those seeking medical treatment suffered from burns typical of exposure to mustard gas, reinforcing that our find of respirator gas masks at our location weren't there for 'Darth Vader' silly rig runs. Little wonder the Kurds were reticent to return home.

It was of no surprise that the Kurds were also wary of trusting us. We were seen to be in cahoots with the Americans of whom they were extremely untrustworthy of after President Bush's previous call to arms. The US had deserted them once, what was to stop them deserting them again? I could totally empathise. We couldn't stay here forever; the Kurds knew that. So how could they be confident they would be safe when returning home? As always the difficult jobs were given to the men on the ground trying to implement the ideas of politicians, strategic plans carried out by the bottom feeders.

The noticeable quiet amongst the lads proved the visit to the camp was a harrowing experience and for the first time as a military person I began to analyse decisions made by those in power. But I was there to undertake the orders that flowed down from above and be professional amongst the carnage that now surrounded me.

What was even more unsettling was our role amongst the confusion of Middle Eastern politics. I'd already sussed that we were here because our closest allies, the US, had thrown a bone to the Kurds then ran away leaving them helpless, so had blood on their hands in the subsequent genocide. But here they were, in front of the world's media, portraying the role of saviours. We were also allies of the Turkish. We'd used the US Airfield in Batman as a transit point and had military echelons in Silopi on the Turkish side of the border also liaising often with Turkish border guards at Çukurca camp. Yet our main source of assistance in Northern Iraq was the Peshmerga. Literally translated as 'those who face death' the Peshmerga are the Knights Templar of the Iraqi Kurds. Strong, fearless mountain militia, they protect the Kurdish name with a ferocity that strikes many with fear. We sat with the Peshmerga often, identifiable by their baggy trousers, jacket and shemagh headwear, to swap intelligence and formulate plans to secure areas. This wouldn't have been an issue if it wasn't for the fact that the Peshmerga were seen as terrorists by the Turks. While the Kurdish separatist movements of Turkey (PKK) and the Iraqis (PUK, KDP) were distinct they were inextricably linked through their aims of a reunified Kurdistan, and supported each other's causes. PKK separatists, for instance, were often harboured on the Iraqi side of the border. A simple explanatory analogy to this dichotomy is to imagine the British allowing the French to use its airfields while training and supporting them, all while the French were plotting operations with the IRA - a strange situation I am sure you will agree.

With such a delicate balancing act, no little wonder then when one of the Bravo Company signalers, George G received a Mention in Dispatches for averting a huge diplomatic incident by defusing an armed stand-off between the Peshmerga and a US Marines detachment. Such small acts undertaken by the youngest and least experienced are often the dividing line between unheralded peace and armed conflict.

'The feeling of being unwanted is the most terrible poverty.'

~ Mother Teresa, Catholic sister

THE KURDS HAD STARTED their descents back to their homes. Day after day they made their way down the hillside like torpid ants to the bridge checkpoint where we offered water and small amounts of food aid including M&Ms that tasted of washing powder. They were the bare-footed walking dead, some stumbling from exhaustion, some already dead tied to the backs of faltering donkeys ferried home for a dignified burial. Weather beaten, the smell of wood smoke and faeces followed them, impregnated in their clothing and skin, ingrained dirt giving them a layer of poverty camouflage. At least they were leaving the camps and returning home. Behind the filth and sickly pallours now shone sparkling eyes of hope.

Our next phase was in action, and I reveled in escorting the IDPs back to their villages, riding shotgun on the cab of the dump trucks that were used as mass taxis for the returning Kurds. It was a hugely humbling experience seeing the Kurds arrive back in their villages. In the silent, empty tumbleweed streets, Iraqi Army destruction surrounded us. Buildings of note now rubble, many houses' pastel muds scorched with the black of ordnance and soot, and pictures of Kurdish leader Jalal Talabani peppered with gunshots by Government troops - the irony being that he would be their future leader. Yet despite the devastation, returning families wept with joy. The threat of future reprisals in such a fractured country was only to be expected but for now they didn't care - they were home. Having been displaced for four months surviving the harshest of mountain winters, many losing loved ones to either the retribution of the Iraqi Army or disease, this was now where they could move forward. Standing there, shaking hands and being hugged by the many men thankful of our assistance was incredibly touching. Although a tiny cog in the huge war machine of the allied forces, I felt I'd done my bit, and I found more satisfaction from assisting in a humanitarian capacity than trying to shoot the fuck out of someone for the sake of our government's interests.

Laughter, as they say, is free medicine and it was my fellow marines' incessant humour that was the saving grace - the release from the harrowing scenes we'd witnessed. A couple of lads stole into the elite 2rep French Foreign Legion's camp and rather than nicking one of their sheep used for fresh rations, they wrapped it in black masking tape writing '40 Cdo' in white mine tape to let the FFL lads see that commando operations were more than just killing people. Sod's operas were organised, frivolous kangaroo courts were played out and games nights were undertaken. It is quite a surreal experience dressing up in a bin liner suit in the middle of Northern Iraq compering horse racing games. I'm convinced not many have done that.

We also had the morale boost of Jock. He was an RMR corporal, a grizzly old timer with a Glaswegian accent so strong we only understood half of what he said. We did understand every ounce of the scream, though, when his head was on fire.

We'd recently moved our position down to the riverbank. Jock had his own two man arctic tent and seemed happy putting it up by himself. It wasn't long before he was yelling as the ants that had covered his tent were now biting him in all the dark places one doesn't want to be bitten. Watching him breakdance in agony was only the beginning of the hilarity. Yanking his tent from the ground, he found that his tent was positioned on top of a small fissure housing a huge ant's nest. Exacting revenge like an Iraqi Tank Commander, using naphtha fuel as an improvised formicide, he poured the fuel down the fissure before setting fire to it, laughing wickedly at his mass killing. Seemingly though, the naphtha flame didn't last long enough so, as it died down, Jock threw more flammable liquid onto the flame. This was an amateur mistake, as the properties of naphtha fuel are such that it is:

a) Ideal for military stoves

b) Extremely, and I mean extremely, flammable.

So flammable in fact, that as Jock poured the liquid onto the flame, the flame followed the liquid back up into the flask from which he poured, setting fire to the flask he held in his hand. In a state of total panic, instead of throwing the flask to the floor like any normal human, he started shaking it about, throwing flaming liquid all over his head and chest. Luckily naphtha burns at a relatively low heat so watching him burn wasn't going to melt him immediately. But we had to act quickly. With the water point at the other side of the position the better option was to dowse the flames with the nearby stack of the high protein 'Ensure' milk shakes that we issued to the Kurds. As a group we opened the cans of milkshakes covering him in strawberry, vanilla and chocolate flavours until the flames were extinguished leaving him looking like a Neapolitan Nikki Lauda. With his injuries slight, it was by far the funniest thing we'd seen on deployment.

What wasn't quite so funny was an operation to recover the bodies of two BBC journalists on the tripoint of Iran, Iraq and Turkey. Although maps pinpointed the border, the reality was that these tribal hinterlands were wholly porous. No one really knew in which country the bodies were. We'd surprisingly been given permission from all three countries to allow British commandos on their soil to extract them. In the bleak inhospitable mountains of former Kurdistan we formed a cordon, while those chosen collected the corpses. We'd been informed a Turkish guide had killed them, other rumours suggested PKK involvement. Whatever the story, the bodies were in a bad state after the local wildlife had been at them.

***

Our task to secure the bridge over the Great Zab complete, we moved west and occupied the impressive hilltop town of Al Amadiyah, a Masada-like fortress that commanded the local area. We assisted the locals in rebuilding infrastructure, carried out patrols and had fantastic haircuts at the local barbers who specialised in the best cutthroat shaves this side of Fleet Street.

We rotated through Al Amadiyah to a small village of Qidish where we occupied what once was a school. Manning the gates we'd get constant complaints from one man who irately asked why his neighbour's lights were brighter than his, or why weren't we fixing the nearby generator. Seeing as though it'd been destroyed and was about as much use as tits on fish, a troop of bootnecks with a tool set consisting of a Normark knife and some masking tape was hardly going to fix it. We also received visits from the local Peshmerga who looked on in confusion as we disarmed them prior to entering camp, offering them a raffle ticket as a receipt for each AK47 handed over. I was more confused as to where we'd managed to get hold of raffle tickets.

The local kids seemed to find us a novelty, and smiling as they did, would always walk away with something, whether it was a handful of sweets or a marker pen tattoo of the Globe and Laurel.

As well as local tattoo artists, part of our job here was to overlook one of Saddam's palaces in Enshki.

While he had ultimate power in his country, the least he could do was to employ a decent architect. Saddam's garishly ostentatious taste was hardly classy, hurting our cultured eyes as we looked upon the monstrosity from the vine covered balcony used as an overt OP. 45 Commando previously manned this OP and had been shot at from the Republican Guard who manned the palace watchtowers. After the brief exchange, the sentries realised taking potshots at bootnecks wasn't worth losing their lives for. It made life much easier for us.

We found the Kurdish people most hospitable. We often were invited to houses for a chat, gaining small snippets of intelligence. There were clearly elements within the area intent on payback, in whatever form, and keeping a close eye on the various actors in the town allowed us to keep a lid on instability. Unfortunately, sometimes payback was given, and on more than one occasion a house had been targeted, their concrete walls brightened with the spray of red, the result of a gunshot to the head. This was the reality of the Kurdish people. Even with our presence, with Saddam at the helm, they would never truly be safe.

Of similar concern was the Troop Officer's regular disappearance to speak with the local Peshmerga. Without a radio, he was becoming a liability and it wouldn't be long before subversives in the area would take advantage of his cavalier, if not well meaning, approach. It came to a head one midnight. I'd only returned from changing over the camp sentries and grateful to get into the comfort of my warm, if not crusty, sleeping bag. The other two troop corporals, Deano and Gaz, were fast asleep.

'Corporal Time, wake the other section commanders, I need to brief you on something,' ordered the troop boss.

Neither was happy. Deano, like me, was a new corporal, Gaz, however, was highly experienced and was awaiting his third stripe and promotion to sergeant.

'Can't it wait until the morning boss?' said Gaz, a man who had seen and done much, was hardly excitable.

'No it's important, get your rig on,' said the Troop Officer.

'Well if it's that important tell us now.' Gaz, by now, had sat up, but his lack of movement was a poker play to get the boss to explain.

'Well I've had some local intelligence that the guardroom at Saddam's palace has possession of a nuclear device.'

Our laughs weren't what he expected to hear.

'You mean that fucking concrete sangar at the front gate?' said Deano.

'Yes.'

'Yeah, that fucking box where the blokes don't even have a fucking torch?' said Gaz.

'Yes.'

'They've got no torches but a nuclear device.'

'Yes.' The evidence wasn't exactly stacking up in the Troop Officer's favour.

'And where did you get this intelligence from?' asked Gaz.

'The local Peshmerga.'

'Who you've been out with I assume. You pissed?'

'We had a couple of drinks, yes.'

Gaz laid back in his bed, 'Fuck off boss. Come back tomorrow.'

'Sorry?'

Gaz bolted upright again. 'You're asking us to go on a half arsed recovery of a nuclear device from an armed force, something that probably needs a bit of expertise, based upon some drunken twat dit spinning and showing off? That palace is fortified to fuck. "Ooh we've got a nuclear bomb, shall we put it underground, safe where no-one can access it, or shall we put it with a couple of fuckwits on the main gate next to the paraffin lamp as we can't afford fucking torches..." Boss, seriously you got to have a word with yourself.'

The slinking away of the Troop Officer suggested he did.

Having three mutinous corporals wasn't an ideal scenario for him so, as a pre-emptive strike, we notified Dixie, the Company 2i/c. Dixie was the only officer I knew who proudly displayed home made tattoos. An extremely switched on bootneck and great leader of men he forged a fearsome hierarchy with the 6'8" CSM. The Company Commander was the figurehead but these two ran the show and were formidable as a team. They both let our Troop Officer know his actions were not what was expected of a Royal Marines officer, and although we had a fractious relationship, I respected the fact he paraded us as a troop and apologised unreservedly for his poor attempt at being a Kurdish equivalent of 'Lawrence of Arabia'.

With little warning, we left Northern Iraq. The briefing gave us enough time to pack, clean the grounds yet leave Kurdish anti-Saddam graffiti - 'Saddam's nob smells' being my British favourite - and be ready for the Puma Helicopters to collect us to go back over the border into Turkey. No time to say farewell and good luck to the locals we'd become close to; just the heavy swirl of debris from the departing helicopters was all the early rising locals would see of our departure.

Our departure was inevitable, but I can only imagine the disappointment of those we left behind. Their feeling of fear now we weren't present to support them and, once again, the anger that those strange men from the West had just disappeared without even the courtesy of a goodbye. Op Shaven Haven was over. Bravo Company was about to be the first to be deployed onto Op Warden.

***

Op Warden was flowered up in importance but was basically a rotation of guard duties, sunbathing and the occasional aerial sortie in US Blackhawk helicopters to enforce the no fly zone. It would make for a long, slow month.

Co-located in a huge tented camp with the Americans, French, Dutch, and the Turkish; our morale was supposed to be raised with an international sports festival. The first volleyball game between Bravo Company and the Turkish deteriorated into a free for all punch up, the Turkish offended by our lycra shorts tucked up our arse cracks and us being offended by them being quite good at volleyball. We did excel at our own concocted sports - 'who can down the most dry coffee granules' and 'speed eating of out-of-date canned fruit' - neither of which was keenly followed by any other nationality.

Our usual nightly highlight became a trip to the huge air conditioned DEFAC - the US military mess hall, that offered the finest food one could hope for in such conditions, complete with ice cream machines and unlimited cans of 'soda'. The deepening blues of the darkening skies blending into the darkness of the land ahead would be the signal for the evening entertainment - the fireflies of tracer rounds and the staccato of automatic gunfire from the PKK Kurdish militia, brothers of the Peshmerga, our ally in Northern Iraq, engaging our Turkish ally's government installations in the nearby town of Silopi. It was hard to feel any support for either side, so we just sat there in our deckchairs, soda in hand, maintaining an air of comfortable neutrality watching the night sky being lit up by feuding forces pointlessly killing each other into yet another stalemate.

The only thing we had to look forward to on Op Warden, apart from the return home, was a weekend R'N'R that had been kindly organised. After a night at the US Air base in Incirlik we'd proceed to the nearby, but unknown, resort of Kiz Kalesi. No worries, we were sure it would be a great weekend.

I love Turkey. Istanbul has to be one of my favourite cities, its energy hard to match. In my younger days my toe was dipped into Turkish culture in the form of loads of British and German holidaymakers drinking themselves into sunstroke in the bars of Olu Deniz that has some of the best beach space the Mediterranean has to offer. Despite contravening obscenity laws, you will see plenty of topless flesh, yet foreigners are rarely arrested for it, although some of the hairy men that stand around certainly should be. So on this visit, after pulling myself out of a 24 hour period where I could not even motivate myself to get out of my camp bed, then suddenly flying into a cerebrally emancipating 10 mile run, I considered myself a bit of a Turkish veteran. Determined to make the trip less beer and beaches when we pulled up in Kiz Kalesi, about 120km west of the Syrian border, I was confident that the resort could tickle my cultural inquisitiveness. Truly relaxed, I looked out to the castle that sat a couple of kilometres out to sea thus renamed the resort 'Rubble In The Oggin' and pondered my future once the tour was to end.

Cultural awareness is an important asset to have when abroad. It was a phrase rarely used back in the early 90's so we were all guilty of blindly assuming that all of Turkey would be the same as somewhere like Bodrum, the party capital of Turkey where I'd holidayed after returning from the Caribbean. We should have noted our geographical location. We were nearer Syria than Greece. The women, far from being Northern Europeans in just cheese string bikini bottoms, were Burqa-clad women buried up to their necks to keep cool from the hot Mediterranean sun. We should have noted the lack of bars, only one on the beach and a couple shouldering the sands. There were no other Western Europeans here, but as I adore visiting places few others visit, it gave me that wonderful trailblazing feeling. Having walked around the small bazaar and smoked a shisher at a café, culturally satiated I felt content. Unbeknown to me at this point a couple of Bravo boys, amphibious experts remember, had managed to sink a hired speedboat and another lad had managed to bite a hole in a lilo that just happened to have the French Ambassador lying on it. We'd been there two hours.

I found a shit shop selling rubbish inflatables, perfect to wear to a beach concert that evening. I managed to throw together a silly rig of my skin-tight speedos, a floppy hat with a ridiculously large brim and a toddler's rubber ring that had an inflated puppy's head smiling inanely at the wearer. That was it. Three items of clothing that would see me through the night. Beer, silly rigged bootnecks, and burqas - what could possibly go wrong?

4 Troop were out in force all dressed in togas fashioned from the hotel bed sheets, and we felt we were truly on holiday as the open aired concert blasted strange music to the growing throng of locals out to enjoy a night of traditional Turkish music and evidently bootnecks who were now dancing badly on the stage. The locals didn't really enjoy our act. We weren't in Bodrum; we were sharing a holiday seemingly with the Taleban version of Club 18-30. Our bare flesh, flaunted at the innocent eyes of the local women, only brought ire from the local men, who voiced their disapproval through the open channel of bottle throwing. Not used to being victim to such actions (after all I'd received a standing ovation for my role as 'Jonah' in the school production when I was 12) I, like the others on stage, responded with humorous ripostes. They didn't find us funny. Within thirty seconds the crowd had turned into a sea of flying fists; locals fighting toga-clad bootnecks and silly men in kids' rubber rings.

The police finally arrived. They smiled, then spoke to us saying, what I can only assume was, 'say hello to my big stick'. In the best traditions of police violence, not seen since the miners' strike, they started whipping anyone dressed in silly outfits, conveniently forgetting to hit any of the Turks who should have been arrested for wearing tank tops, trousers and trainers. Knowing fighting the police wouldn't be our best option we beat a retreat along the sands where I saw one of the RMR lads, Jimbo, ignore this option, fighting off a gun-toting police officer with an oar. Stood nearby watching, with that inebriated stare that suggests a stomach pump is only a shandy away, swayed a toga-clad Spud, casually pissing in the sand.

'You not gonna help him?' I asked Spud as I ran past.

'Never fight with your nob out,' Spud replied.

It is impossible to argue such logic, and advice that I've heeded ever since.

With raw lesions received through police hospitality, we decided to retire, giggling and full of adrenaline, for another big day in this surreal resort only noting at this point that my speedos had been shredded in the mayhem and I was walking about with my knackersack exposed to the world.

Sunlight had only just hit the windowsill when a loud knock on the door woke us from slumber. Unfortunately, it wasn't the normal knock of hotel staff - that tentative rap followed by a polite request to attract attention of those inside. This knock demanded the occupants respond, and quickly. I got up, and without even acknowledging such a threatening bang on the door opened it to see the 6 Troop Sergeant standing there with a few armed police. There was no polite introduction.

'Get up, get your kit together, be in the foyer in ten minutes.' His face wore a plume of anger. 'No, make it five.'

TV cameras greeted us outside. We were being kicked out of the resort. Even the CSM and the Company Commander weren't spared from the rough treatment of the police. With the threat of a diplomatic incident looming, our trip back to USAF Incirlik came with a warning from the CSM to be on our best behavior as the guests of the US military. We tried, honest. Unfortunately a couple of 6 Troop lads decided that, as they'd missed out on the troubles in Kiz Kalesi, they'd get involved in fisticuffs with some US Airmen over something as important as the ice cream vending machine.

As 45 Commando had been kicked out of Incirlik only weeks earlier for joy riding a brand new Hummer vehicle into a lake, little wonder the Americans decided to send us back to Silopi. Given we only had four days of R'n'R, being kicked of two places in two days was quite an achievement.

Our return from the tour was unheralded; the only attention we received at Exeter Airport was a cleaner telling us not to walk on the floor she'd just mopped. There was little news in small wars and the adulation received from larger scale operations was instead filled by the loathing of many Taunton folk, again becoming social pariahs prevented from certain pubs and spat at in the street by those who knew little of what goes on beyond the borders of comparative safety. Thankfully, in more recent times, social acceptance of our forces has seen them welcomed as heroes. The main pity is that horrific casualties have had to be sustained to ignite such support.

***

The upside of being away on deployment was the eight weeks leave graciously offered. Our pockets were full of money, men could try to kill themselves by buying motorbikes that were far too powerful, go mountain climbing, parachuting, or just spend quality time with loved ones. I decided on a holiday to a location where British societal norms were frowned upon in favour of Asian decadence.

There were seven of us, including Carl, Spud, Simmo and Bez, who decided to spend a fortnight in Thailand. It was elongated to four weeks, then ended up being a full eight weeks holiday. We couldn't leave.

Not due to any physical ailment, although Carl spent a few days invalided due to his poor motorbike skills, just our disinclination to return to the banality of suburbia. We were kings of South East Asia. We had no money worries, we could buy whatever we wanted; whether it was a trip to the unspoilt beauty of Ko Samet or hand made suits that we'd pay for at fitting, but never collect as, in truth, they were shit. We could sit, drink '*Singha*' beer or '*Mekhong*' whiskey and eat Pad Thai all day. So we could stay up for days on end, we'd drink '*LipoVit-an*'- industrial strength caffeine syrup pick-me-ups available well before any of the poor substitutes invaded the Western market. On the odd sober occasion, we'd watch and sometimes participate in the Muay Thai boxing matches. If we could organise a Queensbury rules boxing bout we'd invariably win. As soon as we reverted to traditional Thai boxing, the local's ability to destroy bootneck shins meant we would often lose, although Spud would swear to this day that he won the fight despite not being able to walk afterwards. We'd try and be cultural by seeing Buddhist monasteries, then a few hours later undo all the good work with those upstairs by watching zombiefied women doing vagina tricks in dark dingy bars where neon lighting gave the performers a multi-coloured sense of mystery. Some women had the ability to pop balloons with darts squeezed from between their legs. How a woman can place an eel head first up her vagina then manage to squeeze it back out head first is a mystery I still cannot fathom, and while cocooned in a permanent bubble of whiskey fuelled ecstasy I felt no guilt when jumping into a bubble bath wrestling naked women, or jumping up on stage naked as the girl dancers, trying to look sexy with my meat and two veg dangling like a turkey neck handbag containing two quails eggs.

On the hot mornings, we could lay hungover on the beach enjoying lengthy massages limited only by the endurance of the masseur. In the cooling afternoon walking in the monsoon rains, we'd get drenched, so rather than pay 10 baht toilet fee would piss our pants while sat at the outside bars as no-one would notice. The only memory of doing so would be the acrid stench of ammonia the following morning from the dried shorts discarded on the room floor.

It was here that I was first introduced to that novelty that is so common in this part of the world - the 'kathoey'. Known otherwise as a 'ladyboy', 'he/she' 'Kythai', 'lassbloke', girlgadgey' et al, these marvels of transsexuality are generally accepted in Thai society. Indeed it has been reported that in one of Bangkok's most revered universities a third toilet for transsexuals has been built (how it works I have no idea), but in Western culture they still polarise opinion.

For servicemen, from post war Bugis Street, Singapore to modern day Sitang Camp, Brunei; kathoeys have a tenuous connection. Both may have members who feel outcasts to normal society leading them to choose their followed paths, others revel in the exotic. The sexual nature of many a serviceman, may also lead them to see a kathoey as a challenge to take on, rather than something abhorrent. Of course, many servicemen have been caught out thinking they had pulled themselves a 'girl' only to find a hand full of man giblets when things were getting steamy. As for those who jumped in with both feet, I wonder if having sex with a kathoey would actually be against military law? It wasn't confirmed homosexuality, especially if the kathoey had undergone the full 'cut and tuck' operation. I'm sure the powers that be knew of the shenanigans that went on. After all, they'd also hear the baritone voice at Sitang

Camp call for "good times for soldier boys" from a specimen that looked like a miner who had fallen headfirst into a cosmetic counter. They could not fail to hear the stories from the lads who, not caring whether the 'girls' still had their twig and berries, would be quite happy to indulge fully, and for the more depraved would be generous and give the kathoey a 'reach around'.

It is easy to be fooled by the modern day kathoey. They can now access some of the most advanced plastic surgery and gender reassignment technology available. However, those with less available funds, the results may not be so rewarding. At the lower end of the aesthetic scale some kathoeys wear an ill-fitting wig and have the appearance of an androgynous air steward hit by a lipstick covered boxing glove. At the opposite end of the spectrum some are absolutely stunning, their deportment straight from the finest Swiss finishing school. More feminine than many natural women, only their hand size, sometimes a pronounced Adam's apple, their skeletal structure or their voice can give them away as someone who, while wholly feminine, are not wholly female. To the uninitiated, a successful kathoey looks a beautiful woman, and the more advanced gender reassignment surgery becomes, gender lines become even more blurred. On my last trip to Thailand I had to look twice when propositioned on an errand to buy soft shell crab flavoured crisps for the kids. She truly was a goddess, and only her aggressive manner told me 'she' was a 'he'. Saying that, the meek table host at our local restaurant looked like a badly tanned Ricky Gervais, much to the confusion of my 11 year old daughter.

'All animals are equal but some are more equal than others.'

~ *'Animal Farm'*, George Orwell

IT IS DIFFICULT TO RE-ENGAGE with military regime after eight weeks of cavorting freely in South East Asia. While I was gutted to be back, I was relishing the challenges that work could throw at me. Unfortunately, all there was to look forward to was black shod exercises in such exotic locations as Salisbury Plain and Sennybridge.

Luckily, the Intelligence Cell was expanding. With only a marine under the Intelligence Officer and a sergeant, the recent Northern Iraq deployment had shown a larger cell was needed to optimise the unit's intelligence capability. The Intelligence Marine at the time was Dolly, with whom I shared a mess on the Caribbean deployment. He'd done a sterling job on his own but was discovering pastures new. He put a good word in for me, so when interviewed for the job the cell already had some background on me. I would hope so; they were supposedly military intelligence.

I must have impressed the panel as I got the job ahead of the other candidates. Perhaps my train spotter ability to name nearly every capital city in the world impressed them, or my newly acquired cultural awareness borne from ignorance of prancing around half naked in the more conservative areas of Turkey. Neither would help win a firefight but intelligence wasn't about shooting people. It was about turning information into intelligence so others could shoot the right people.

I was in my element and fitted in immediately with the cell. Led by the Int Officer who, at first glance, I'd have disliked - an ex public schoolboy who tried to advocate the pleasures of fox hunting. Upon getting to know him better my ridiculous stereotyping couldn't have been more wrong. He was a fantastic man, intelligent (one would hope so), a lateral thinker, and despite his background, understood the workings of a council estate bootneck. The new Int Sgt was a man who made me laugh at every opportunity. Typical as a PW in his application that everything had to be just right, he was a great leader. Dizzy, the leading hand matelot - the unit 'phot' and absolute star and master of the camera, who could make me look 6 feet 3 and good looking on a polaroid. The Int Cell had the unit Illustrator attached to it. A unique branch in the Armed forces, a Commando Illustrator was a Blue Peter producer's wet dream; a man who could knock up anything from a sign for the MT Office to a perfectly scaled 3D map of Dartmoor from a few loo rolls and some sticky back plastic. Joining me would be Jonty and Knocker, two marines from different companies who would complete the Int Cell. Together, the three of us would be sent on a basic intelligence course where we had the choice of Bulford (army garrison town full of dirty pongos and smack bang in the middle of Salisbury Plain) or York (student town full of congenial women in the middle of one of the best nights out in the UK). After a long debate, we chose to go to York.

Imphal Barrack in York is usually a bootneck-free zone. Rarely do the normal inhabitants of 2Div HQ get to host 'Royal', and so we played up to this fact by ensuring we entered the galley (never a 'cookhouse' even in pongo land) in arrowhead fashion. I was fortunate that both Jonty and Knocker were big, good looking bastards which took the heat off me being 'rats' and the army girls sat at the various tables must have let go a little bit of wee as we attacked the hotplate, taking our one sausage and one slice of bacon that had the mandatory circle of gristle embedded within it.

The course itself didn't cause us any problems even if we did make the most of York's infamous hospitality. Even with the previous night's beer swilling around in my gut, my 'Marvo the memory man' abilities held me in good stead. Jonty, Knocker and I could recognise, and recite the characteristics of a ZSU/234 or a BRDM far better than any of the other students who'd sat in their bunks at night trying to get to grips with the orbats of the Soviet Army. I had a birthday while there so a very messy night led to me following birthday tradition by commando crawling across the dance floor with a bottle of cheap champagne up my anus, down two flights of stairs and onto the wet pavement, just to prove what a twat I could be when drunk. A corporal from an army regiment I can't recall, couldn't understand how we could go out every night, drink until we could breathe no more, come back in the early hours of the morning (they noticed the time as we bumped from one bed to another like pinballs in the accommodation) and still get the top three marks the next morning of the revision tests. The only rationale we could respond with was the unconscious bias of saying, 'we are bootnecks'.

Just by being bootnecks meant we had the extra-social capital where our commando flashes suggested we were expected to do better so had to outdo any other arm of the services with whom we happened to be with. On this occasion, we'd earlier sussed our cerebral advantage over the rest, therefore, we found no satisfaction in getting higher marks, so by self inflicting mental anchors to our brains through getting rat-arsed gave it a more even playing field, just so we could boast we had come top of the course *and* been shiters all the way through.

Hungry for more courses, I left behind the unit once more and headed for further intelligence courses including an advanced Soviet Studies course at the old Army Intelligence School in Ashford, Kent. My Intelligence Officer had a struggle to get me loaded onto the course as it was aimed for officers; but fighting my corner as a good Royal Marines officer does, managed to secure me a place.

Looking at the others subscribed onto the course, I was the youngest by a country mile, and on the shoulders of my fellow students, I saw more pips than the inside of a value orange. Amongst them was the course officer of my Basic Intelligence Course in York who was slightly curious as to how a lowly corporal could be sharing the same classroom. Indeed, I was the only non-commissioned rank and felt somewhat lonely eating my fatty steak and kidney pie dinner while the rest enjoyed afternoon tea in the ante room of the officers' mess. I did, however, get an invite to the course Team Leader's house.

On exchange from the US Air Force, the colonel held a drinks evening on the first night to welcome us all. She could have not been over five feet tall, her thick spectacles and tree trunk legs reminiscent of 'Velma' from '*Scooby Doo*', and it did cross my mind to trip her over just to see her crawl along the ground whining, 'My glasses, I can't see without my glasses!'

It would've been easy to be overwhelmed, being the sole non-commissioned rank, the only one without a regimental or Oxbridge tie, and happily alone in not wearing a V-neck sweater. But without the protection of uniform and overt rank, I stood comfortably to assess the scene around me. I listened intently on the colonel's chats within the hubbub of genteel conversation. I soon realised I was her intellectual peer. She, on the other hand, would never be able to do a 30 miler or carry weight. She wouldn't even be able to pass a BFT wearing boots for fuck's sake. As I stood people watching, I fathomed I could hold my own against anyone around me, even if I'd never previously eaten a canapé. It would be easy to be arrogant amongst those who did not wear the green beret. Arrogance is a negative trait yet I, like many a bootneck when in the company of strangers, tried to walk the right side of the divide that separates arrogance from the value of self worth. As a human, I faltered occasionally over to the wrong side, humility forgotten over a misdirected show of cockiness, but self-awareness would kick me back, usually when in the company of my colleagues from the Corps. Yet here I intended to show that, despite the best intentions of the British class system, when given a level playing field, council house scum could hold their own against the social elite - something that the Corps, in part, recognised in its application of officers. Then again, I was glad not to be a commissioned army officer if what I was now witnessing was a cross section of their existence. The choice of wallpaper was the only interesting thing one could draw from the evening. Even my old course officer turned into a bland, homogenous goo of sycophantic uniformity. The evening was as enthralling as a bowl of salad, all the ingredients trying to be the nice cherry tomato, but just fading into another characterless piece of soggy lettuce, without enough substance to challenge anyone. Taking each person as a separate entity, should one slip away no-one would actually mind or even notice. I wondered whether they often hankered to join the lower ranks, even just once, to savour what real life was like; break loose and try to plait a woman's hair with their feet, set free the shackles of convention and run naked and free down the high street or piss their bed and actually be able to admit it to their colleagues; just once. It would do them good.

With being in the company of commissioned officers up to the rank of full colonel and a couple of MI5 geezers, I had a harder time of rising to the top. These were intelligent people; all successful in their own worlds, not average soldiers, but leaders, managers, strategists, policy makers, and Pimms drinkers. So it was even sweeter when I got the top marks in the final exam. Not that I would ever get to utilise my skills properly, the course was designed for high rankers, but satisfaction prevailed from just being able to wink inwardly at myself and know that I could excel in such esteemed company. The possibility some would secretly be pissed off by being beaten by a bootneck corporal just made my success even sweeter.

Unfortunately, riding upon the shoulder of my success sat my dark twin, its malevolence praying on any weakness shown. The palette I often carried around to paint my world with vivid colours would be ripped from my hand, replaced by a tar brush to blacken the world I had to suffer. As normal, it sprang when I least expected it and upon going to Newquay for another weekend of frivolities it crushed my head as if a grape. I felt it creep up on me mid-afternoon as I was packing. Pete, Frankie and Dinger were most excited about the upcoming weekend of surfing, supping and shagging. The only 3 'S's that were in my mind were sleep, solace, and self-harm. Again the excuse of a hangover from the previous Thursday night activities was the excuse to my quiet and my sleeping en route I hoped would rid me of a large portion of guilt. Yet despite the laughs that surrounded me, I ended up slinking from a pub at 8pm on the Friday evening. A vacuous black sucked the spirit from me, my head pressure cooked within the faint, yet unshakeable buzzing that filled my throat with that unmistakable urge to cry. The white light at the end of the tunnel when death is resisted is replaced by the urge to step into the dark foreboding abyss - an escape route for those who can no longer bear their turmoil. I was tempted to walk towards it, yet thankfully became immobile. Selfishness writhes inside when in such a place, so thought not once about the confusion on the other side of my hotel room door where, on the Saturday morning after they'd knocked, I could here my best mates talk about me. Only their repeated door rapping knocked me from the temporary grave of my bed, where the entombment of rumination left me wondering where I belonged in the world.

I finally awoke Sunday afternoon parched from not drinking since Friday night - water is purely optional when the world hates you so. It was Sunday evening, my mind still dizzy, when I found the courage to step into the light of the outside world. Frankie, Pete and Dinger were all confused - and annoyed. These were the days before mobile phones so, despite me being plums most of my life, they believed my story of meeting a girl and spending the weekend involved in a sexual marathon. I even pretended she was essence, which probably pushed the boundaries of believability.

Lying about these episodes was now becoming as natural as expecting them. As before, I could pretend to feel a little ill and tell the Int Cell personnel, who would check why I hadn't turned to, that I had reported sick and had been given bed rest. I reported sick a lot, but I was never once ill. Physically I was in my prime, but the darkness that occasionally compressed my skull immobilised me as if in a straight jacket. Guilt, self-conscience, ill confidence, anger, all attacked me through the nights of sleeplessness, with no answers or the ability to relax as my escape to the torment. With no reasoning the dark twin would flee, laughing at me with a warning that he would return whenever he wanted. To forget him I would again get out my box of colours and paint my world the brightest it could be.

**

As the Intelligence Corporal, I thought my time would be spent undertaking more exciting pastimes than reading the newspapers. In the days before the internet, open source information was digested through the broadsheets and various intelligence reports. I was slightly naïve to think that intelligence work would be action packed. It took a studious and patient mind to crunch the information accrued and churn it out the other end as relevant intelligence and it took me a while to learn the finer points of being an intelligence operative. The forthcoming Norway deployment, where everything moves a lot slower, would be the perfect environment to learn.

There is an old saying in the Royal Marines "If you can fight and survive in Norway you can fight anywhere."

I'd been in the Corps for five years, a corporal and had never been to Norway. Some, especially the arctic foxes of 42 or 45 Commando, would say I wasn't a proper bootneck. I could live with that. It wasn't somewhere I was longing to go. Apart from the amazing fjords there was little to muster my interest to visit, and as for being there in the winter, I would have certainly been happier with a temporary swap draft to 'RMR Honolulu'. But it was a rite of passage. The Royal

Marines, in my time, were guardians of NATO's Northern flank, to be first response to the threat of those pesky Soviets creeping over from Murmansk into Finland then into the UK's back yard - Norway. With this responsibility it was only natural for every bootneck to be arctic trained. 40 Commando was in its second winter deployment, having skived off it for many years until some bright spark decided we should swap factor 2 sun cream for ski wax.

I prepared for the conditions by travelling around the humid heat of South East Asia for Christmas leave, so surprisingly enjoyed my Arctic Warfare Course (AWT). Yes it was cold, but a different cold to what I'd witnessed before. This was a numb cold, one you knew could kill, and thus respected. I never shivered once as we were taught how to operate so effectively in the extreme cold. The devil was in the detail and such seemingly rudimentary tasks as clothing ourselves properly to prevent both over heating or getting too cold would keep us not just alive, but fighting fit. The Mountain Leaders (MLs) who taught the AWT were fine, knowledgeable, men who in true bootneck instructional style could be ferocious one minute then the next make you laugh until masking tape was required to hold your sides together. Even when we fucked up, the subsequent beastings were taken on the chin. This environment would kill and if we chose not to get it right, then we'd suffer the worst of consequences.

I thought having being skiing three times on holiday I'd be a bit of a Franz Klammer. So too did the HQ company storeman. Normal downhill skis come in many sizes and while technology has advanced sufficiently for skis to be virtually bespoke fitted, in those days, were sized according to your height. Usually a novice skier would be advised to get a ski that came up to their eyebrow, which in my case would be around a 165-170cm ski. The civilian downhill skis I'd bought were 175's as I fancied myself as an intermediate level skier and could handle the extra length. Military skis (known as Pusser's planks) come in two sizes: 190's and 210's - basically size 'very long' and size 'very fucking long'. The bindings were far different to the ones I owned. Downhill ski bindings are solid blocks that hold the ski boot tightly giving slight lateral movement to shift weight. The bindings on military skis looked like a slinky designed to wrap around the small groove in the back of our specially designed military ski boots made by a cobbler who only had the hardest of leather to work with, fashioning them as drill boots but with added stiffness skillfully sewn into the upper. The free heel design of the binding and boot allowed the military skier to langlauf, or cross-country ski similar to a skating gait.

The storeman handed over my 210's with a smile. 'You've skied before Mark, you'll be alright. Plus we've run out of 190's.'

So as I studied the skis that seemed to reach the sky, I realised that the similarity between my sleek black civilian carbon fibre downhill skis and the big fuck off bastards that I now struggled to hold was about as negligible as finding commonality between Mike Tyson and Stephen Hawking.

My proficiency on my own bespoke fitted skis was also in stark contrast to my ability on military skis. In fairness, I'd never gone downhill humping a small house in the shape of a bergan on my back. The manicured tourist ski slopes of the Alps are brushed, rolled and flattened each morning unlike the wilds of Norway where snow lays where it falls and coats the ground in a blindingly white topography reflecting the ground below. When skiing in formation, the lead man blazes the tracks that others try to keep within. This does compact the snow slightly, but the natural unevenness and varying depths caused by wind ensured that when a novice skier, such as I, slid slightly off the ski trail, they would fall as if been shot. I'd usually fall over the front of my skis with the full weight of the bergan flung over my head if I was lucky or, more painfully, into the back of my skull. The extra weight would thoughtfully pile drive my head even further into the icy floor, my abuse given to every single bit of equipment currently worn was muffled by snow and frozen grass that filled my mouth. The only solution was to breath deeply, count to three and attempt a heave to get the 60kg weight from my body before trying to right myself. With my legs sliding like a baby ostrich, I'd attempt to stand while my skis were eager to continue without me. Once up and settled with another deep breath that chilled the lungs, off I would tentatively ski once more, only for it to happen the next time I hit anything that could be exaggeratingly called 'a bump'.

Norway, as expected, was all rather dull. The scenery was stunning; however, with one click of a meteorological finger the clear blue skies could be engulfed by the curdling of heavy clouds and the crisp linen whites of the sleek mountains could be hidden within the thick woollen blanket of a snowstorm, desensitising the body's compass as well as exposed skin. The soldiering element for me was rather bland. Skiing or being driven in the back of a BV tracked vehicle to another location where we would dig a huge hole, pitch a tent for the night, then pull pole, to do the same thing day after day wasn't my idea of fun. The cold environment didn't bother me, my natural impatience found the whole slowness of operating in such conditions quite toilsome. Luckily, we had most weekends off, fantastic for a bit of downhill skiing and a lot of cross-country skiing races where others in the company would ski off as though quicksilver flowed through their veins. I, on the other hand, struggled along as if ballasted down with broken breezeblocks with a less than silky gait that suggested someone had sneakily tied my bootlaces together. If I appeared on '*Give us a Clue*' Lionel Blair would have thought I was miming the film '*The Jerk*' or the lesser known book '*The Irrefutable Misplacement Of A Sunshine Commando*'.

The local towns were expensive for imbibing activities, so the pillaging of '*Appelcorn*' spirits from the camp bar served as an aperitif, leaving guys unable to walk even to the four tonner transport into town, often donning silly rig. While through the week we wore thick arctic clothing to protect us from the elements, by the weekend we had acclimatised enough to wear sheer net body stockings, lycra boob tubes and bin bag suits. On them were pinned the mandatory reflector disc to prevent being run over when laying drunk and frozen on a roadside and, apparently, protect us from the severe cold.

When on camp we still had to do guard duties. Of course, Muggins here caught up the role of Guard Commander when the whole of the Brigade had a senior non commissioned officers (SNCOs) regimental dinner. The Corps RSM would be in attendance and so too the many SNCOs from around the Corps so, as well as trying to purge the three months of garlic powder I'd inadvertently dropped into a drunken pot mess, I had the added pressure of babysitting inebriated SNCOs all the while ensuring the guard stayed focused on the task at hand.

As in most regimental dinners, endless bottles of wine beer, port, and rum were consumed and as the night drew down, many SNCOs were still fighting the urge to retire by drinking their own body weight in strong alcohol. Just to add insult to injury, instead of the lads being able to sleep when not out on perimeter guard patrol, as was normal procedure, they were 'volunteered' to help clean up the regimental dinner's wake of destruction. At 2200hrs guys don't really mind. At 0200 it becomes a bit of a pain and blokes tend to drip like a septic cock. At around this time, Dave 'Kamikaze' Carswell, the duty signaler, reported that a local Norwegian, living in the nearby collection of houses, had contacted him stating he had heard a cry for help from the deathly bleakness outside. The perimeter patrol was dispatched to the area where the noise was supposedly coming from. I took with me George G, fresh from receiving his Mention in Dispatches for his service in Northern Iraq, to sweep the area hoping the fresh air would offer temporary relief from the constant stench of garlic that I breathed, farted and permeated through every pore.

It was one of those cold, still nights where temperatures dictate that death is only a foolish act away. George G and I were wrapped up well, mittens and wooly hats donned to protect the extremities from the dangerously low temperatures. We heard the feeble cry simultaneously. As we hadn't done cry recognition studies, to us it sounded like an ill kitten, yet followed the sound as it repeated. With head torches on full beam we saw a hole in the ground. Peering inside, two feet below us, was a near-unconscious colour sergeant, recognisable from the rank on his formal mess dress. He was curled up in a foetal position, his breathing worryingly shallow. It didn't take the brains of Bamber Gascoigne to see he was in serious trouble. We dragged him out and cradled him between us back to the camp bar. Ordering a sleeping bag and the guard to stop clearing up, I contacted the Duty medic. All I could do was get him out of his frozen clothes, and get into the sleeping bag with him to give him some body warmth. It was like cuddling an ice statue, one that minded not that I reeked of garlic. Within a minute I too was shivering uncontrollably. The next few minutes the guard took turns jumping in to the garlic filled bag next to him. None of us felt weird for cuddling this balding old man even if it did look rather dodgy. We couldn't afford to warm him too quickly, and he couldn't drink the warm sweet tea we offered. His death did cross my mind. His breathing wasn't improving, and what little he did breath, was just alcohol fumes that added to the seriousness of his condition. His hands were gnarled like a broken old tree and his feet cadaverous. Within ten minutes, he was under the care of the Norwegian paramedics. If we hadn't yet respected the conditions of Norway, we had now. We had a quick debrief chaired by the Duty Officer and despite clearly saving a man's life, the guard's reward was to return to washing the dishes of the regimental dinner.

As a result of his drunken wandering into the wilds of a Norwegian winter dressed in a formal shirt, dickie bow, a light mess jacket and twill trousers finished off with a nice satin cummerbund, the colour sergeant survived, but lost a few toes and fingers from severe frostbite. Any aspirations he may have becoming a rhythm guitarist with 'The 'Shadows' were clearly dashed.

It appeared it wasn't his first experiment with hypothermia. He'd fallen over the side of a landing craft earlier in his career and had to be fished out of the icy waters of northern Norway. Some people have all the luck.

> *'Sex is not the answer. Sex is the question*
> *Yes is the answer.'*
>
> ~*Swami X, comedian*

AS PROMISED, I SPENT EASTER leave, in Thailand to attend an infamous wretch's wedding. The faint-inducing heat of kneeling for eight hours in the sweat-box of a monastery did prepare me for the upcoming trip back to the tropics of Central America.

A small company group had been tasked to go to Belize for a six week deployment to do some jungle training and assist the army infantry regiment out there. Fortunately, they required an intelligence element and who better to do intelligence than someone who had previously been there? Even better for me, Pete my best oppo from the previous Belize trip, was going. In truth there wasn't anyone else to do the trip. Knocker and the Int Sergeant were going away to assist Bravo Company in the Caribbean and Jonty was being punished for nicking an ambulance - fear not, there was no patient inside - that would be irresponsible. It was clear Jonty was not cut out for a career as a paramedic, so with the Corps preferring to see future potential over past discretions, they allowed him to apply for a commission, where he would flourish to become one of the highest ranking officers in the Royal Marines to start his career as a humble marine.

There is a lot to be said for not returning to former glories. I was now stuck in Airport camp with little time to spend out and about due to my commitments. Little had changed. Airport Camp still stunk of the Belikin beer factory effluent, and the paint on the front door of Raoul's Rose Garden still peeled in forlornness at still being out of bounds to service personnel.

Our final task of the trip was a remote boat patrol that was a welcome break for me. I'd eventually managed to escape the ops room to spend the final week away cruising up and down the rivers in a rigid raider. I reveled in these low impact patrols - hearts and minds exercises where we would make contact with the locals in remote villages, gather any intelligence, and treat any illness. We stayed at an indigenous locals' camp on the third night and shared their home brew. Whether it was lost in translation I don't know, but it seemed we were drinking fermented diesel. How they extracted the alcohol from diesel was lost on me but I woke in the morning with ulcers covering my mouth and a hangover that could only be described as 'death wishingly painful'. I wasn't the only one poisoned, and it was decided that the patrol couldn't continue with us all feeling so shite. We had completed most of the tasks already and saw no point in carrying on one final night in our communal states. Besides, we had a long weekend to wrap up the trip. The order was given to the coxswain to break the engine on the boat. He was a little confused. He was a Royal Engineer amphibious specialist attached to take us. It was his first time with bootnecks and as this sort of thing was our bread and butter was quite perturbed that he was being ordered to break equipment. Pete, the Patrol Commander explained his reasons, and so whatever the coxswain did, worked. We had no working craft. Radioing in to tell of our disappointment, we recovered the stricken vessel and returned to Airport Camp a day and a half early.

Pete and I hadn't really decided what we were to do for the long weekend, but now we had a day longer we could venture further afield. Without a destination in mind, we ventured to the civilian airport, despite being told not to go abroad which, we reasoned, was an unspoken order to actually go abroad. I think.

We agreed to get on the next flight available and as we entered the small departures hall. We looked up and smiled as the next flight stood shining like a beacon.

***

You know when you are near the foot of the social ladder when you leave the airport in a courtesy minibus full of people who, unlike yourself, actually have reservations at a hotel of repute.

You've got eleven stops until you get off.

Ten stops for the others to alight at the 'Sheraton', the 'Marriott', the "Sorry Sir, I regret to inform you that you are declined entry on the grounds of you being scum," and all the other hotels that 'welcome' its customers with a revolving door guarded jealously by some twat in an outfit borrowed from that 'Buttons' geezer who was invariably played in pantomime by chirpy cockney type Charlie 'Awight my Darlin' Drake.

Stop nine takes you out of the main tourist zone.

Stop ten takes you into the ghettos. People become inversely proportionate to the amount of garbage on the street. Those that are brave enough to walk the cracked sidewalks give a glazed stare of confusion as to why a bus would travel down this particular road.

The bus driver, whose previously colourful face had diluted to a deathly pallor, took us to our stop.

His final piece of advice was, 'Mind how you go, be careful, and keep to the main road.'

'Cheers,' we said, like co-joined twins.

'No problem. Welcome to New Orleans,' he said as the pneumatic doors closed on our second thoughts.

We'd arrived a couple of hours earlier on a flight by the El Salvadorian airline TACA, known as Take-A-

Chance-Airways. Our flight had started in Belize City. For those not really in the know about Central American flight times (you need to get out more), the dura-duration of the flight is 1 hour 50 minutes. Those that have seen our in-flight film 'The Hunt For Red October' will possibly know that it is of a similar length. Those who, more surprisingly, managed to stay awake will definitely know it's 1 hour and 20 minutes of pure tedium with the final half an hour making the film worthwhile renting for a Monday evening as it comes with a free ice cream. Therefore, when the aircraft switched off the film on the 20 minutes to land signal, the film had just managed to gurgle to life, and Sean Connery had just started to kick some ass. The plane erupted in mayhem as if some nervous paramilitary had sprayed a rainbow of 5.56mm rounds into the cabin. The Captain had obviously heard the commotion from the rear and tried to appease the passengers with regurgitated safety messages. It calmed the hordes only slightly. Eventually after a term of one-way negotiation, there followed an air of uneasy silence, as if the flight deck had informed all that there was something not quite right with the workings of the aircraft. No, this was the calm before the storm where Connery fans were waiting until the plane landed before executing the cabin crew if they didn't restart the film.

On taxiing, the bi-lingual announcement hesitantly requested that the passengers remember to take all their baggage with them. Pete and I became instant losers of the impromptu 'musical statues' contest. No one moved; they wanted to see ol' Seany boy and his toupee hand out a serious spanking. They had been glued to the screen watching visual mogadon for so long, they now wanted the ecstasy. We just managed to get to the exit as the passengers turned violent. Men, women and children erupted, as the film screen remained blank. An air stewardess was tossed aside as she tried to stop Mr. and Mrs. Irate of 26 Anger Street Belmopan from crucifying the Captain. A sharp exit was the better outcome for the two of us before we became hostages of the first hijacking in the name of mediocre film endings.

Out of the melee, we endured the rigorous customs check of arriving from Central America. Knitting needle prods purposely punctured all my puncturables left me with a carrier bag leaking shampoo, toothpaste and shower gel through the hot, sweaty concourse.

To choose a hotel, seek advice from an information desk, not the airport information board, as it's a lottery. Needless to say, we went for the latter and like all lotteries we came away losers. It'd make life a lot easier for the independent traveller if they listed these hotels in league table format because we then would have ignored the South Devon Sunday Second XI league - sponsored by Poopascoop - that our chosen hotel would have been fighting relegation in. It looked nice enough. Not the cheapest, but in a comfortable price bracket. We should have smelled a rat, (and later we literally did), when the line wouldn't connect when trying to book our room, but half the fun of travelling is not knowing what's around the corner.

After stepping off the bus, we tilted our heads confused but sure that the building in front of us was the one we'd booked. On the airport information board the picture showed a rather unpretentious, yet welcoming building where the Waltons could possibly have lived. In real life, it looked like something from the cover of 'Drug Den Weekly'. Pete and I could only sigh, shrug and simultaneously look at each other á la Seventies jesters Mike and Bernie Winters as we journeyed into the grimy abyss. We opened the rickety outer mesh door. Thousands of dried crusty insect corpses, their dead wings like microscopic leaves, desecrated the mesh obscuring the view inside.

When I was about 8 years old, I'd often go to the park to play football and get covered in shite. The following school day, I'd play football at every playtime, furthering my covering of shite. Returning from school, I'd go straight out to play football, get covered in shite, etc. etc. This routine would go on for about a week before my gran would even consider asking me to change clothes. The smell of those stained kneecaps after day five was a cross breed of piss-infested damp underlay, and dog food. This is how our new landlord smelled. He stood like a stunned rabbit amongst the garbage of the porch. He would have been an ideal example to set for small children who had never voluntarily washed behind their ears. His hygiene left a considerable amount to be desired and it would have taken a great amount for him to be desirable.

It was 6pm and the night called. One night in 'L'Hotel D'Underlay et Provision de Chien' couldn't be so bad - but it was. Rubbing ourselves with the fallen damp wallpaper was more refreshing than the shower that was on strike and would yell 'scab' at the basin taps that trickled a pathetic stream similar to that of someone suffering prostate problems.

After changing our room, which, oddly enough, was pretty easy as we were the only guests; we managed to freshen up to take on the task that lay before us - the taming of the French Quarter.

In a country where it's easier to get crack cocaine after 2am than a Miller Lite, it's odd to find a city where supping beer, falling down, and licking your stomach lining from the gutter is common place - 24hrs a day.

By following the crowds we invariably found ourselves in Bourbon Street. Made famous by a street lamp, it is so much more. Here in the Deep South of the USA you have a street named after a chocolate biscuit when it should be named after the God of pissing up (whatever his name is).

As a result, we found ourselves venturing as far as anywhere with a 'bar' sign outside, proclaiming there would be no attempts at pulling women, just the consumption of beer, mainly to overcome the fear of sleeping back at the Munsters' set. I can't really remember much about the first night. My last recollection was talking to a strapping leather-clad chap who seemed very hairy but very friendly, inviting me to a bar where he'd show me some good ol' Louisiana hospitality. Pete did his best not to offend this poor bloke by pulling me away from the door of this darkened cellar bar full of sweaty men dancing to Erasure.

New Orleans tourists usually get up at 7:30am have a hearty breakfast to set them up for a fun filled day of touristy things. We woke at 2pm, piss-wet through with sweat, then left the hotel without delay to spend a fun filled day looking for a room. I do recall saying half the fun was not knowing what was round the next corner? Forget that. Believe me there is little worse than slobbering in high humidity heat in beer sodden clothes, stinking like your ex landlord, trailing shampoo, toothpaste and shower gel around dusty streets looking for somewhere to rest your head. Plenty of hotels yes, but on this particular day there were plenty of hotels that had all of a sudden just become fully booked as these two fuckwits stumbled through their door.

By 6pm we'd drunk the town dry of decaf in a misguided attempt at self re-hydration, so now we were bad breathed fuckwits. Luckily we managed to step into a Fuckwits-R-Us hotel on Canal Street, which was surprisingly posh. Set in the usual French style, we were welcomed along parquet floors that echoed to give exaggerated space to a small lobby where the French-accented receptionist welcomed us quizzically.

'Do you have rooms spare?' we asked parched, sweaty and flustered.

She must have sensed our desperation, 'Yes we have a twin room, but it's only available tonight.'

'How much is it?' I asked hoping not to sound like I was a cheap skate.

'$125,' she answered nonchalantly.

'OK,' Pete answered. It was a lot for a pair of scruffy arsed bootnecks but we had little alternative.

'Per person.'

We looked at each other.

'Ah fuck it, yeah we'll have it,' I said probably in a way she wasn't expecting, one night of luxury was well deserved.

Once allocated, we were disappointed to find the room was as spartan as King Leonidas' contemporary research and development laboratory. The concierge did have the courtesy though, to find us a hotel for the next two nights. So, with our lodgings sorted, and a big hole in our communal pocket, it was with renewed vigour that we set out on our second night around the Quarter. We felt lucky, but not in a punky sort of way.

Twenty minutes later, out of a list of 100 feelings, lucky was somewhere around 99. I was kicking a wall repeatedly outside the bank where the cash machine had just swallowed my card, (1st of those feelings was excruciating pain as my toe had bust in my anger-induced kicking frenzy followed by 97 expletive ad-verbs). Pete had reached the daily limit on his debit card after withdrawing $260 to buy a Noel Edmunds jumper that he'd seen in a small designer shop whilst drinking coffee. We had $35 between us for the night. $35 doesn't go far, especially when a certain idiot tells the barman to keep the change from the $10 given when the first round only came to $5.50. I think Pete slightly lost faith in my powers of money management at that point as he rightly pointed out that it was only 8pm, and we had around $10 each left to party the night away with. He also rightly pointed out that I was a complete twat. There would be no repeat of the previous night.

We sat in Pat O'Brien's like two sad blokes on a 'Cheers' set at the bar just so it would be easier to spend our fortune. Obviously the puppy dog eyes that we now sold to people around the bar worked, as before long two women appeared from nowhere with one thing on their minds - unfortunately that thing was us. I've no qualms about meeting with a girl with a severe weight problem, after all it's what is inside that counts. Even if the outside has ginger hair, or a hair lip, or a squint. But put them all together, add a tattoo classily placed on the breast, then it was with great relief that this particular girl latched straight onto Pete. The girl that I spoke to was more aesthetically pleasing - her tattoos were spelt correctly. What did catch our attention though, were their bulging purses, which contained more notes than a slow typist's in-tray.

Rich women = heaven.

Rich, ugly women = heaven with some reservations about where it was all going to lead.

Nevertheless, they were paying for the drinks and the conversation did flow after the mandatory stupid American question, 'Are you Australian?'

We set her straight.

'Oh, English? Do you live near London? I've got a relative in the Outer Hebrides - David Andrews. Do you know him?'

'Andy Andrews? Oh yeah I know Andy.'

'You know him?!'

'Yeah, pity about him dying like that.'

'Dead? Oh my God!'

At this point I had to explain my sarcasm before she phoned Scotland to console her Great-granddad's-second-cousin's-niece's-husband's-step-sister.

More money than sense is a commonly used phrase in the United Kingdom. Over there it should accompany certain people's details on personal documents such as their driving licenses.

Conversations should never go like this:

'So what do you do in England?'

'We're both elephant exterminators on the Yorkshire Moors.'

'You're what?'

'Elephant exterminators, you know, kill elephants,' Pete echoed matter-of-factly.

'Oh geez, that's awful, what makes y'all do that?'

I then nodded my head in mock agreement as she then proceeded to rant on about the ivory trade and the threat of extinction caused by people such as myself.

'Well, I agree in principle to what you are saying, and elephants do need a certain amount of protection, but do you know what caused the most deaths in young women and children in the Yorkshire countryside last year?'

She paused - it was hardly a question she was likely to answer.

'I'll tell you…' I added another pause to build the anticipation. 'Tusk impalement, that's what.'

'Tusk impalement?'

'Yep. Elephants get a load of good press but you ain't seen 'em when they smell milk. They proper love milk they do. They go banzai and end up stampeding through these small villages where the dairy farmers live.'

'Wow, really?'

'Yeah, really. This is where we step in, we monitor their movements and as soon as we get wind of any trouble we go straight to the villages and protect the people.'

Pete decided to take over the mantle for added authenticity, 'It's quite scary; we have to make a human shield between the villagers and the stampedes. If the elephants come too close we have to shoot them. Sometimes it takes 10 or 11 shots to bring the big ones down by that time they're only a couple of yards away.'

I jumped back in, 'Yeah, but you can take down a calf with one good head shot.'

'Oh yeah, of course a good head shot will kill a baby,' Pete added motioning brains being blown away.

'Last year, Pete was awarded the 'Elephant Cross' for gallantry. He single-handedly culled a herd of elephant calves who had gone berserk and stampeded through a dairy farm looking for strawberry milk.'

'Oh you're so brave.'

Pete's eyebrows rose slightly as his partner pressed his crotch. He was in. Official.

So was I.

The film should have been titled '*Dumb & Dumber & These Two*'.

Two hours later, via various fast food outlets where Pete's 'girlfriend' amazed us with her amazing eating ability; we were in their apartment. Pete was less enthusiastic and quite dismayed at the situation, but had now resigned himself to the fact that he'd have an over enthusiastic eating machine jumping on his bones in a short while.

In the morning, I woke up aghast. When she woke up I thought she still had the pillow stuck to her head. She didn't. Thinking back to that hairy nipple, then looking to the floor where her sanitary towel lay stuck inside a pair of stained knickers, I was thankful I'd fallen asleep while fumbling around her folds. I decided it was only fair to come up with a poor excuse to leave. I could just have as easily said that we were about to join the local dyslexic Devil worshipping sect and so sell our souls to Santa, but settled for being late for a pre booked tour of a local alligator farm. Stuffing pastries into our over-dry mouths, we left as quickly as our taxi would let us, which at least they paid for, back to our $250 an hour hotel, via the bank to retrieve my bank card.

Pete, by now, had realised that the purchase of a $260 jumper wasn't such a great idea after all, and asked me to buy into it. I questioned the profitability of shares in a pullover and decided against an aggressive market takeover.

Bourbon Street had drawn us back by 3pm. The strip bars were plying their trade as usual, having both done the rounds before we knew that we would have to take out a small mortgage similar to one that could buy a three-bed semi in rural Hertfordshire, so returned to Pat O'Brien's to chill and to feast upon traditional New Orleans fare. We were now hooked on gumbo, a stew-like soup that gave us our daily sustenance leaving us piss that smelled, not of *Sugar Puffs*, but kippers. We could also plan our night ahead which, if it went according to plan, would mean we meet two rich and charming divorcees, have a foursome, get married, get divorced, get half their dosh, then live forever on a private island. Sorted.

By 9pm, part one of our plan had worked. We *had* met two charming women who *were* rich, well richer than us. God loved us. A few hours later I knew God had changed his mind and didn't particularly like me at all.

Pete hadn't impressed his girl. I, on the other hand, felt like Ron Jeremy. The girl sat with me had my flies open at the bar, yanking my todger as if she was pulling a turnip out of the ground while suggesting we partake in illegal acts of depravity. For the first time ever, I'd managed to attract a girl who wasn't interested in Pete. Luckily she didn't have x-ray spectacles, as she would have seen his tackle belonged to a prize-winning racehorse unlike mine that belonged to a self-conscious pygmy shrew. Not withstanding the handicap of inadequate tackle, I felt the burgeoning excitement of carnal pleasure. Therefore, it came to the decision: do I go with this sex-starved strumpet, or go with my mate? Being a real good friend I decided to leave Pete.

On his own.

At Midnight.

In New Orleans, the murder capital of the Deep South.

He couldn't believe it. I gave him $40 as a token of my gratitude but asked he stay out of the hotel room until 6am. I did actually feel a modicum of guilt as I saw him forlornly drag his heels and bottom lip along the sidewalk.

These ladies were in New Orleans on a short vacation from Alabama. They were staying at a relative's house in the city suburbs, and being the gent that I am, decided we should all travel by taxi to my hotel via the relative's house to drop off the other girl. It was their first night in town so they were a little confused and drunk when we alighted by a bridge spanning a small canal a little short of our destination. They tried to get their bearings before leading off into the area on our side of the canal. It was of no great surprise they realised we were heading in the wrong direction. As we turned, I caught sight of three figures emerge from the shadows and casually remarked that we were going to get mugged, to which the girls began to laugh. The laughing stopped abruptly when the middle of these three guys stopped us in our tracks clearly high as a kite. I wouldn't have been too bothered if this guy had pulled a knife, or even a pistol, but when he pulled from within his coat a Heckler & Koch MP5K sub machine gun I expected to be another gangland killing statistic. He threw my new 'girlfriend' against the wall, thrusting the stubby muzzle down her throat like a gun-obsessed dentist. Blood dripped from her ears after the mugger ripped the earrings from her ears and yanked away her necklace.

'Leave her alone,' screamed the other girl. 'Here, have my cigarettes.'

I looked at her with incredulity. 'Oh right they're going to be satisfied with 20 Marlboro lights,' I said, probably more sarcastically than necessary. It didn't help matters. I turned to the muggers. 'I've $250 in my wallet fellas,' I piped. Really I had about $30 taxi money, but it was a gamble worth taking. 'You can take it all apart from these, they're no good to you,' I added, enthusiastically waving my bank card and my mutilated Forces ID card in the darkness.

They seemed suitably satisfied as they scurried, like Hyenas to a carcass, to retrieve my wallet that I'd thrown as far as I could behind them. As they turned tail I grabbed the girls to run, inwardly giggling on adrenaline, back to the bridge. The pseudo bravery given by alcohol may have its detractors, but that night it saved my life.

Once successfully home the girl, who less than an hour before was making me blush with comments I'd rather not disclose to my Aunt Shirley, was shaking like a shitting dog and definitely not in the mood for a bit of rumpy pumpy. The cops came a few hours later, one leaving us safe in the knowledge that he'd been shot in the back on that same street corner the previous year, showing the scar as if to try and impress us.

'If you'd gone further along the road, you girls would have been raped and killed,' he said. I pulled a fake scared face to try and lighten the mood. He turned to me, rudely pointing his finger. 'And you would have been executed.'

My stupid face straightened. It was only natural to upon such news.

'So you're lucky to still be alive.'

Lucky by now was definitely 100th out of a 100 feelings. I'd been mugged, lost $30 owing another $40, denied sex, and would probably get a kicking from my best mate for leaving him in on his own. All for a bit of filth.

It was about 5:45am by the time the police dropped me back at the hotel, just in time for my head to hit the pillow and for the doorbell to ring. It was Pete.

I answered the door ready to adopt the foetus position. To my surprise he was smiling and had decided not to stove my head in, as he had a woman of his own. Out of nowhere he'd met the winner of 'Miss Heaven' competition and she'd come down to Earth for a night of gratuitous sex.

'Could you and your woman please vacate this room? I would now like to use it, ' he said smugly.

'She's not here, we got mugged.'

'Good. You deserve it.'

'No, seriously. I've nowhere to go.'

'I'm sure I said that to you last night.'

With a little negotiating, we managed to come to an arrangement where Pete would pick me up and throw me into the walk-in wardrobe by the bathroom. There, I would sleep upside down while he had lots of rude fun with the stunning air stewardess.

The girl who I'd failed to touch intimately awoke me mid-morning. Amazingly, despite her near-death experience only a few hours earlier, she'd come to the hotel ready to carry on from where we'd left off prior to our impromptu rendezvous with New Orleans' most anti-smoking gunmen. Now I don't know if you've ever slept upside down in a walk-in wardrobe that at 40°C has no ventilation, with a hangover and chronic diarrhoea. It doesn't do much for the libido. Subsequently, no matter how hard she tried to impress me with what she could do with various toiletries, I could not rise to the occasion. She at least satisfied herself before leaving, leaking shampoo, toothpaste, and shower gel. As she left, sleep was the only thing I desired. As most losers do on holiday, we awoke as it was time to go out on the town.

Back in our two lucky chairs we were picked from the meat shelf within ten minutes, this time by a mother and daughter tag team. All sorts of permutations crossed our devious minds but after a whole night of heavy hinting and double entendres, group sex was declined in favour of separate bedrooms. Pete shuffled off with the mother. I left with the daughter who looked unerringly like Marty Feldman with a wig.

Upon entering a hotel bedroom I expect to find a few things lying on the bed: melted chocolate, a dressing gown perhaps, or takeaway food carton carelessly discarded forgotten in the rush to get out on the town. What I didn't expect to find was a bloke the size of a yak, spread-eagled on the only bed in the room.

'Err, who is that?' I thought it only reasonable to ask.

'That's Jim,' she replied.

'OK, so why is Jim in your bed?'

'He's my brother.'

I thought it advisable at this point to confirm whether she had the correct number of limbs, fingers and eyes.

'He's ill, so couldn't go out tonight.'

It was getting weirder by the second, especially when she insisted on sex next to him.

'You don't fancy going to the bathroom or somewhere else?' I suggested.

'Bathroom? Ooh no that's gross.'

This girl liked her sex next to her brother and only the small space the width of a gymnast's beam would suffice on a wrought iron bed in serious need of oiling.

I could, in the best traditions of gutter journalism, make my excuses and leave, or attempt the world's most careful coitus in fear of death by jealous brother. When you're a bootneck in such circumstances, it's easy to bypass the agonies of indecision...

*'Taking a new step, uttering a new word,
is what people fear most.'*

~ *'Crime and Punishment', Fyodor Dostoyevsky*

I HAD FOUND MY VOCATION in the Intelligence
Cell, yet knew the time would soon be upon me where I
would be drafted from 40 Commando. A draft usually
lasts two years so I'd been fortunate to spend so long in
Taunton. My courses had allowed me to return to '40'
and thus start a new draft but now as a general duties
corporal I was at risk of being sent somewhere I didn't
really want to go. I looked at the options available and
saw there wasn't much in the way of deployments
around the brigade that interested me. I thought time
away from a commando unit may be in order. I'd been
at '40' continuously for nearly five years, other than
time away for courses, so taking up the old adage of 'a
change is as good as a rest' I looked outside of the
brigade for a palette cleanser that would satisfy my
wanderlust.

I saw it. A trip that would visit Freetown in Sierra
Leone, Ascension Island, Recife and Rio De Janeiro in
Brazil, Montevideo in Uruguay, the Falkland Islands,
around Cape Horn, up the western coast of South
America calling in Valparaiso in Chile, through the
Panama Canal before having a two week stop in
Tenerife in August. Had I been looking at a Cunard
cruise brochure? No, but close. This was the itinerary
for HMS Cardiff who was in need of a Royal Marines
Detachment Corporal. I submitted my chit and as ships
drafts weren't usually popular, was immediately accepted. I had six months to wait to join the ship but would
be sent on my Pre Embarkation Training (PET) straight

after summer leave. Going on ship, I hoped, would not only be a life raft for my liver, but also sort out my finances. I wouldn't have to pay food and accommodation charges and would get extra money as I qualified for Long Service at Sea Bonus. Although the ship would be calling in some of the most exotic ports in the world, in between it was hard to throw money around while at sea. I was a 22 year old single corporal. While my salary wouldn't see me driving a Ferrari (or a Mini Metro for that matter - I hadn't yet learned to drive), with only food and accommodation automatically deducted from my pay, I had a fantastic disposable income. However, I was a *bon vivant,* living a pop star's lifestyle on a pop delivery man's wage, permanently riding on a povercraft, skimming my way over the sea of excess en route to the port of Skintsville. I was wildly overdrawn. By the last week of the month, I would always be nervous whether the cash machine would spit out cash or laugh at me telling me to come back when I had some money. I'd little in the way of belongings; the most expensive item I owned was a Sambuca-covered stereo. I did see myself as a blithe paragon of financial profligacy, and most of it was due to hard partying and my unquenchable thirst for travel.

The end of the month was usually spent buying guard duties or selling assets to purchase beer tokens, and other party animals would sell stereos, watches or kit for below market value to the more sensible men at the lifestyle stock exchange of the NAAFI notice board.

My circumstances weren't driven by alcohol, indeed even when I look back through more mature eyes, I still cannot say I was an alcoholic - I had no dependence on it. Those battered by the wet kipper of suburban monotony may deem our consistent nights out as rather self-destructing and sad, yet on camp what was the alternative for a single male - sit and absorb the drivel of banal mind-numbing TV? Waste away playing shit computer games? Save some cash for that rainy day rather than fritter it away on the sunshine of now? Books were usually read on deployments and many lads could consume a novel within a week. We would do our phys, we could do some extra curricular work and we could stimulate the grey matter, but as we weren't 12 years old, a few evening hours were still available to kill, so we purposely chose to go down the pub to fulfill our lives as recycled teenagers, reinforcing our bond as mates, fine tuning our communication skills, seeking alternatives to how self-righteous couch commandos expected us to act. Moreover, while our nights out weren't driven by beer, the offering of alcohol was the lowest hanging fruit when ashore. We were social butterflies standing together, who saw the world as a challenge to take on with verve and gusto with imbibing just a by-product. Drinking did allow further access into a Pandora's box of excess and risk taking to a point where I'd lost the fucking lid, but I never felt I had to drink to enjoy myself. I woke up many times vowing I'd drunk my last drink, but never woke wondering where my next drink would be. I could go out and drink soft drinks all night, and many times I did, but alcohol was the path of least resistance.

It was also a good time to move on socially. My closest mates were being drafted their separate ways due to the peripatetic nature of military life. Simmo was going to Diego Garcia, a terrible draft where for a whole year he would stay on a tropical Island in the middle of the Indian Ocean, taking on the part time role of Customs Officer and 'Brit Bar' doorman, with eight other bootnecks and a glut of military personnel in search of a year's partying. 'DG' was a draft that everyone wanted and was as rare as hen's teeth, so we were chuffed if not slightly envious of Simmo's leaving. Carl was getting promoted, and Bez would be off to spend time at Her Majesty's pleasure within the walls of Colchester military prison for dishing out some barrack room justice to a prick who'd eaten his own bodyweight in disgraceful conduct. Dinger had gone on to Pilot's training, Nick was in 45 Commando, so only Pete remained at '40' thus I became even closer to him. My involvement in Bravo Company's legendary grot parties had petered out and I, like a terminally ill patient with a one-way ticket to Switzerland, felt content that I was ready to move on. While I felt settled in Taunton, there was always a feeling that my permanent yearning for travel and excitement was partly a search for a true home and calling.

My dark twin came and went as it always did interjecting my *joie de vivre* with maudlin through a quagmire of emotional porridge. My quiet times were my own and the deceit of my excuses was hidden by my irascible responses to those who enquired a little too personally. Of course, those who asked most often were those who cared the greatest and my curt responses only deepened their hurt. I hoped that a new start could quell this confusion that hampered me so.

***

My time with '40' nearly over, my last summer leave consisted of spending time with the recently released Bez. I'd categorically stated that I wouldn't return to Thailand. I'd spent my last three periods of leave there and could see it was taking over my life as it had many other bootnecks. As a dose of holiday methadone, we took a quick trip to Magaluf along with Brum who, by his own admission, was a Thailand addict. His last trip had turned sour, as his Thai bride-to-be had done a runner with a wealthy Swiss pervert. We assumed somewhere different would be a great tonic for him. We couldn't have chosen a worse place.

Magaluf was a place where civvies crammed their year into two weeks of madly partying, trying to emulate the 24/7 life of a bootneck. However, we certainly didn't fit in here amongst the throngs of kebab eating youths. We didn't take drugs, we didn't listen or dance to acid house music and we certainly didn't have long straggly hair that seemed to be in vogue. From the first moment we arrived, we hated it. We tried to talk to people but found we had little commonality. The only guy we spoke to for longer than a minute was someone who claimed he was a Royal Marine. Our ears pricked up when we heard him boasting to a couple of women. On further listening we realised he wasn't a bootneck. The lingo he used wasn't bootneck, and the way he boasted of being a commando certainly wasn't bootneck. We politely asked to expand on his profession as we pretended to be eager to join the Royal Marines. He told me I was too small to join, which I found hilarious and allowed him to get a shovel and dig a hole big enough to swallow him and his family. Once we admitted we were actually real bootnecks he froze in horror. Apologising profusely, he admitted it was his ruse to attract women. What made it even sadder was that he was actually in the army - a Royal Welsh Fusilier, and so should have been proud to serve his regiment not try and hide it under a ridiculous façade.

We were military oil drifting around in this saltwater sea of civilian partying. Our shit dancing routines, regularly applauded by local girls back in Taunton was seen as a bit 'weird' and our oft-heralded sarcasm left a few girls utterly bemused. Even my pantomime camel silly rig drew sneers from women when I slapped them with my beer-saturated tail. It seemed society was moving in a direction we'd not yet found, so Brum reverted to type and led us to a whorehouse, reinforcing him as someone who had forgotten how to chat to a lady without first paying a fee. While Brum went off to try and chat up his prostitute, Bez and I sat at the bar drinking over priced local beer, and realised we were truly part of a lost tribe drifting slowly from societal norms entrenched in our own sense of morality, one that was at odds with the generation pushing through. We were on the fringes, and we loved it.

***

A new draft brings with it a new adventure, so I arrived at RM Poole keen to get as much out of the PET as possible. I met two lads, Bill and Paul McGuigan, who would be drafted onto the ship a few weeks ahead of me, so were happy to give me the run down of what to expect on the PET.

There wasn't much to the PET, as expected. A couple of trips to HMS Raleigh in Plymouth were planned, one to undertake a small boat coxswain's course out at the picturesque Jupiter Point; the other a basic seamanship course to learn essential naval skills such as how to throw a hawser and eat endless cheese toasties.

Trips to Portsmouth saw me on an immensely enjoyable firefighting course and a sea survival programme that saw me jumping into a lake then climb into a life raft to assimilate awaiting rescue. It was pretty realistic, although in reality I hoped I wouldn't have to sit for ages, looking like a mutant orange in a fluorescent drysuit, bored out of my tree listening to baby matelots whine on about how they were cold and wet, and how shit the NAAFI beer was, and how bad it was you couldn't smoke in a life raft and how it was rough and how…

Joining ship the week before Christmas leave, I felt like Elvis in his iconic GI film footage. There I was, with my hair gelled back, kitbag over my shoulder, as I thought I should look 'proper jack', to be led down to the seaman's mess.

'Where's the bootneck barracks?' I asked. 'All detachments have bootneck barracks.'

'Not on HMS 'Can Do' Cardiff, Royal,' said the matelot.

The Ship's Captain had decreed we weren't actually a detachment but part of ship's company so would be split up between the seaman's and stoker's messes. No bootneck barracks meant no naked press up competitions, no shitting in each other's pillowcases and certainly no homoerotic oil wrestling. I'd be lost for things to do.

I took the many squeaky metal steps down into the bowels of 3P Mess. With two small square spaces surrounded by 57 tightly squeezed triple bunks I didn't know whether to laugh, cry or just accept I'd be smelling other people's beer, egg and curry farts for the next 18 months. Each set of bunks was separated by a small thin space known as a 'gulch', and I was placed at the far end gulch, a rather exclusive development of beds reserved for the more senior sailors and a couple of hairy arsed bootnecks.

The current Royal Marines (non) detachment was in a state of flux. As we were being drip fed into the ship those we replaced left, changing the dynamic of the group instantly. The corporal I replaced was a crusty old sea dog of many years served. I was the opposite, but even though I was initially mistaken for a young marine, all welcomed me and immediately I felt again the affinity with the sailors as I had on HMS Intrepid. Unfortunately, I didn't have the same sort of relationship with the senior rates. I was given a 'part of ship' led by a rather camp petty officer (PO) who didn't really know what to do with me. Shoving me over the side to do a morning's chipping and painting in a crane-slung bucket with a Wren who didn't want to crack her nails, I realised this ship's draft was hardly the life of a commando. On the upside, I was due to fly to Colorado on the Friday night before Christmas leave with Pete and Nosher, a lad I'd known for years and who was about to embark on his SBS career, which surprised everybody - no one could see how his massive head would fit through a submarine hatch. I only had to wait for my passport to get transferred from RM Poole to the ship. With the efficiency of the RM Poole clerk, Royal Mail and the HM Forces mail service combined, why would I worry that it wasn't yet here?

I purposely didn't parade on the flight deck as per normal sailor routine the next morning. I had no interest in another day's chipping and painting. My claim of not knowing about presenting on the flight deck was accepted, if not believed, and my PO realised immediately he had a problem on his hands. The Detachment Sergeant Major (He was allowed to be the DSM despite no detachment being allowed) told me he would find something more suitable after leave.

Being the final week prior to leave, my part of ship PO allowed me to just get myself sorted with the other jobs that my position would entail. Seemingly my job was sitting in the Dockyard NAAFI drinking tea and eating toasties with the gunners and seamen with whom I was attached to on the upper deck guns. Despite my inexperience of a type 42 destroyer's gun systems I was given the job of Missile Gun Director (Visual) basically controlling all the upper deck weapons including the 4.5" gun and the Sea Dart. Me? In charge of all that armament? I'd been on a week's course at HMS Cambridge to induct me into the job, yet it was a wholly inefficient lead-in to actually doing it operationally and I struggled to understand the morning's start up procedures and even more worryingly the processes involved when actually firing.

Morning checks would be last thing on my mind in a few days when whistling down the slopes of Steamboat Springs, Colorado, I just had to receive my passport. By Thursday night I was worried. I still was not in possession of it. Friday morning would be my last chance, so took myself off to the dockyard mail sorting office to personally search for the right envelope. My passport was anywhere but in my possession.

I was supposed to be flying to the USA for a dream skiing holiday and I couldn't go due to pusser contracting a fucking snail with a limp to deliver my passport. *Fuck*.

The rest of the day was one of ignoring questions of why I was so down. I slunk off to my bunk and wondered what the fuck I was now going to do for Christmas. My mum was now separated and living in the Turkish Republic of Northern Cyprus cleaning a drug lord's huge mansion perched on the cliffs overlooking Kyrenia. I would have been loathe to spend Christmas with her in any case, so with more default mode than planning, took the train back to Taunton via a night in London dressed as the front half of my trusty pantomime camel, and scrounged a bed in an empty Norton Manor Camp where everyone had gone on Christmas leave.

Christmas is a time for family, for being with loved ones and to celebrate all that is good in humanity. I was alone without anyone and the only celebrations I had were the miniature chocolates I'd bought from the NAAFI as they were on special offer along with my *Chicken and Mushroom Pot Noodle*.

It was Saturday 19 December. I should have been on a Denver-bound plane with powder runs, moguls, and desperate housewives on my mind, yet the likelihood would be my Christmas lunch would be courtesy of the Salvation Army. I phoned Bez and told him of my predicament. He invited me up to his house for Christmas but I couldn't impose. I was infamous as a 40 orphan but turning up on someone's doorstep for Christmas was taking it a little too far. I decided to spend the first week of leave with Bez then try my luck with family for Christmas day. My only family option was to call my uncle Jack. My surrogate father, he'd nurtured me through my school holidays while giving my grandparents a rest from my hyperactivity. A renowned Physical Training Officer within HM Prison Service; in my childhood he would take me to the borstal gym where he worked to teach me gymnastics and the fundamentals of weight training. He was the man who taught me not only to swim, but also the man who saved me from drowning when I jumped into the deep end thinking I was David Wilkie, only to realise that being able to do two strokes didn't really constitute me being an Olympic swimmer.

Jack was more than happy for me to join him and the family for Christmas and a dinner plate would be set aside in my honour. With a week of Newcastle Brown Ale followed by a few days with cherished family I was at last sorted for my leave. Pete had left me a key prior to him jetting off to Colorado with a promise of a T shirt so on my return from my Uncle Jack's slid into a few days of heavy drinking with the few bootnecks who now counted Taunton as their real home.

Carl was also here. He'd been kicked out of home the first week of leave and the only roof he could sleep under was his room on camp. He spent Christmas day on his own eating a packet of *Rich Tea* biscuits and drowning his sorrows with the many bottles of beer left by the lads in the 'Albert Fish' bar, his grot's recreation area named after the 1920's serial killing cannibal. I thought I had it bad.

My first course of action post leave was to go on my ships diver's acquaint. With Christmas leave excess still swilling in my stomach I didn't expect to find the physical aspect of the one day initiation at Horsea Island quite so easy and happy that I would be on a diver's course in February. My second course of action was to wreak revenge upon one of the matelots killicks, called Burt who had continued to give me shit about my aborted holiday. I actually liked him, but like all jokes that are repeated *ad infinitum*, it became annoying to a degree where cessation was necessary. The threat of mindless violence was not my style, so I required something with a little more guile to get my own back. He'd started gloating about the girl he was now going out with. Going out with was a bit of a loose term. In the days before the internet, random girls would write random letters to random ships in the hope of finding random love. Burt had picked out a letter from a girl from Sunderland, whose photograph he now waved in my face. She was beautiful. Actually, no, she was mega essence. How could Burt, the ugliest man in NATO, manage to pull her? She'd obviously not yet seen a picture of him. And so being the kind hearted matchmaking Eros that I am, took a photo of him at his worst, then sent her a letter together with a photo of me looking my bronzed best. It was a bastard trick of the highest order. But I thought it funny.

It wasn't long before I got a reply from the Sunderland girl. She was shocked I could do such a bad thing. As pious as she was, admitted I was far better looking and Burt did indeed look like a shaved duck on steroids. She would continue to write to me and drop Burt like a hot stone. Good job she wasn't shallow, then. I made it my business over the next few weeks to ask Burt about how he and his girlfriend were getting on. He was confused as to why she'd stopped writing, despite him writing to her frequently. My evil plan had worked, and although I should have, I felt not an ounce of remorse.

I not only had my eyes on a pay rise through diving pay, but on one of the Wrens on board. Although Sunderland girl was good correspondence, the oft-coy glance with this particular Wren gave me feelings I'd not yet encountered - it certainly wasn't indigestion. I first spoke to her with the chat up line, 'What's wrong with your finger?' As she scratched the dermatitis that irritated her hands - love rarely starts more romantic than that.

Her name was Jo and I'd first noticed her dressed in foul weather gear on the flight deck throwing steel hawsers. She was an intoxicating mix of frangipanis and marine diesel and had the most wonderful eyes but the world's worst shaped beret. If she could stir my primordial mating instincts in such aesthetically unpleasing clothes there was certainly chemistry - I'd have to have a word about her headdress, though.

Job wise, I'd been politely kicked off my normal part of ship and put under the tutelage of a Gunner PO called AJ, an affable man, he immediately named me Corporal Thalidomide, putting me in charge of the small arms store. For someone who'd little interest in anything nautical, this was my perfect hideaway. AJ left me to my own devices, as long as ship's armoury weapons were kept pristine. I'd turn to when and where I wanted, usually with the rest of the bootnecks. Our only allowable time together was our compulsory PT that we'd stretch out just long enough to make our bosses question whether we were plotting mutiny. I'd take stand easy whenever I chose and finished when I thought the day's work was complete - usually around 1130hrs.

At around 1131hrs one morning, all my South American Fantasia dreams were erased in one swift meeting. The Captain, who had important news, called the whole ship's company together for a clear lower deck.

'In light of the ongoing Balkans conflict,' he began. I didn't like the preamble I have to say. 'Our planned trip to the Falkland Islands has been cancelled. Instead we are to redeploy to the Adriatic Sea as part of the NATO Operation 'Maritime Guard', to enforce the UN embargoes on vessels transiting to or from the conflict-ravaged Former Yugoslavia.'

I should have been disappointed; after all, instead of visiting many exotic places where women wore dental floss as bikinis, we'd be going where women wore body armour. Yet I was keen to expand my operational worth and while I could say goodbye to sun cream and 'I've been to Rio' T-shirts and say hello to cam cream and anti flash uniform, I could at least undertake specialised operations. As a consolation I still had my ships diver's course to crack so could get an operational dive under my belt to boot. Just to kick me in the nut sac, my diving course was cancelled. We were to deploy a.s.a.p. My well anticipated pay rise went down the same plughole as the dreams of wasting it on Impanema beach.

*'The direct use of force is such a poor solution to any problem,
it is generally employed only by small children and large nations.'*

~ *David Friedman libertarian theorist*

WITH THE OPERATIONAL BEAT UP complete, the Adriatic tour turned out to be as fulfilling as eating a large cake. As the Royal Marines on board, we were kept as busy as a cat burying shit due to our main job of boarding and searching vessels suspected of carrying arms, munitions or anything else on the NATO naughty list.

RIB (Rigid Inflatable Boat) launched boardings could see us skimming along a glass-like surface reflecting the deep blue skies towards something as aesthetically benign as one of the many small Lebanese barges that scuttled across the Adriatic carrying cattle in conditions horrific enough to send even the most bloodthirsty carnivore reaching for tofu. Access was easy in these floating bovine morgues even if standing knee high in viscous cow shit wasn't. Most days; however, we'd speed towards our target in great RIB-throwing, sky-hiding swells; the sea spraying our drenched, rattled bodies. We'd enjoy great air time whilst speeding mercilessly to supertankers that became mere models within the demonic greys of an angry sea. On such large swells, to correctly time the jump onto the target vessel's pilot rope ladder was akin to appearing on 'It's a Knockout'. We did find it comedic, if only to overcome the danger of our situation. Even with the RIB turned into the target vessel, a poorly-timed jump coinciding with an inopportune swell could crush a man between vessels, so every time one of us successfully grabbed a piece of rope, we momentarily breathed a sigh of relief before shouting well humoured shit should they have failed to get on the pilot ladder without a steely glint.

The next challenge was to actually climb the pilot ladder. A veritable smorgasbord of design and quality, a ladder could range from a steady, well-constructed object with a stabilising rod, up to scaling a couple lengths of frayed string with a few rotting bits of dowelling pushed through. Climbing a few metres wasn't a major issue. However, if it was a gargantuan cargo ship that loomed over us, access hatches were up to 25 metres above the water line. The salt-sting would be washed from our eyes by the blinding rain as we jumped laden with webbing, weapons and a daysack full of search equipment. The ascent up saturated manila twine, pirouetting like whirling dervishes by gales smashing us against the metallic shield of the riveted hull revealed the reason we did so much rope work in basic training, and why they tell you 'don't look down'.

Alternatively, we'd be delivered to the target by Lynx helicopter, a far quicker and drier ride. As the first one to disembark from the Lynx, I'd sit high above the angry sea with 60 feet of rope on my lap. At the target point I'd drop the rope, don my thick suede fast roping gloves, and slide the 60 feet to the deck below, keeping the rope close to my body, spreading my legs to allow maximum descent speed then braking a few feet from the deck with a Chinese burn to the rope. We could dispatch the eight man boarding team within ten seconds, which not only was impressive, but looked really good for hot chicks and film crews.

As sod's law tends to strike often with me, the only time any TV crew of note took footage of us fast roping was onto a certain Russian tanker a couple of hundred metres from HMS Cardiff. The conditions were horrific. Storm force winds meant that the heavy rope, which usually just dangled perpendicular, was blowing at a 45° angle over the side of the ship. If the helicopter was buffeted by the high winds or the ship moved in the heavy swell I could slide 20 metres down the rope then a further 20 metres down the side of the hull straight into the swirling wake of the tanker, which would have made a hilarious excerpt on one of those home video shows but a pretty nasty death. Fortunately, I didn't go over the side to certain death, I landed on the tanker. In a snotty heap.

Never before had I fallen over on landing, but the distraction of stardom and high winds took hold as I reached the rust red deck. The abrasive paint finish common to such craft was conspicuous by its absence. This tanker's painting team obviously preferred a veneered finish to their deck so with the underfoot friction of coconut butter, I slipped heavily onto my coccyx. Yelping with pain, I rolled over into a run to take up a position to secure the area, but again slipped on a couple of metal steps, yet again falling, causing my arse bone to smash the sharp corner of each tread. Hitting it once was bad enough. Falling on it a second time was not only clumsy, but also excruciatingly painful. Nevertheless, I had a job to do, so limped up to the crew's quarters to coordinate the search teams, all the while trying not to vomit with pain.

If the physical suffering wasn't bad enough, I became the laughing stock of the ship; my fall witnessed by the many who'd been keen to assist the TV crew. My dignity was as bruised as my arse, and Daisy, the ship's PO medic, gave me a couple of days bed rest to allow the pain and swelling to subside. But laying in bed is a rather tedious pastime, especially when all the lads are boarding ships, lifting weights on the quarterdeck or running in circles around the upper deck getting dizzy after a run of anything more than a lap. My wisdom questionable, I challenged Bill to another session of phys. The hangar off limits due to Lynx maintenance, we used the tiny compartment deep in the ship's hull that seconded as the Royal Marines store and rudimentary gym consisting of a punch bag and a rowing machine - the physical equivalent of a toolbox consisting of a hammer and a roll of masking tape. Ignoring the pain sitting on the small saddle seat, I rowed as hard as I could while Bill punched hell out of the bag. We continually swapped exercise until sweat flew from our bodies. My brush with secondary school physics ended with an 'O' level B grade so realised the roll of the ship could assist in the momentum of certain moving objects. One of those objects was the saddle seat on the rowing machine. As I pulled forward my bottom, as usual, came away slightly from the seat giving me the ability to thrust quickly backwards. Usually, in a more serene environment, the seat remained under my arse. However, the ship violently rolled - at the perfect time to slide the seat forward without me on it, so as I pushed back to take another row my already excruciatingly painful coccyx came crashing down not on the seat, but on the thin crossbar it slid upon. Bill's laugh turned to concern as I gurned in agony. Pain had been a constant companion during basic training but agony of such gravity was alien to me. Should a nurse have asked me to rate my coccyx pain out of 10 I would have said 13. If being skinned and wrestled in chicken salt was an option I'd have voluntarily participated. Numb from the arse down, I feared a severe injury. I lay on the floor dazed and nauseous with pain, trying not to move. Finally, I felt my toes, then my legs, and gradually rolled onto my side to be helped up the ladder by the now chuckling Bill to the decks above where I immediately went back to Daisy the PO medic. His only option was to order me to bed with a charge of disobeying a direct order should I decide I wanted to again be an international standard dickhead. This time I heeded his warnings, despite all the other bootnecks calling me 'weak', 'mince', and 'gaylord'.

***

In between boardings, my time was mainly spent high up on the upperdeck as MGD (V) with my GPMG gunner Ant, who joined the ship just prior to the deployment with Bruce. Two lads of similar age, similar ambition and would take similar paths in their career. Bruce would end up at Hereford distinguishing himself in Afghanistan and Ant would see even more glory in the SBS, yet here they were pretty green marines but keen as mustard to do well. I was blessed with the lads I had to work with. The ten bootnecks all had something to offer, even Monty. Monty had been nicknamed 'Snap' by the matelots due to his admission of only knowing how to play one card game. Monty didn't do himself any favours. He wasn't the sharpest tool in the shed, yet was gold for the detachment. He was fearless, and would offer assistance to anyone. While he stuffed up on many occasions, his morale value was unquestionable. However, when he did stuff up he stuffed up badly.

Stationed around the ship were big fuck off red buttons, only to be pressed for the express emergency of someone falling overboard. We'd been told this on our induction, but evidently Monty registered only the one we'd actually been shown. Surely only a buffoon would have to be shown every single overboard emergency button? We hadn't been shown the one on the quarterdeck and Monty, in a moment of immense buffoonery, allowed his curiosity to get the better of him. Rather than asking what the big fuck off red button was for, he just stood transfixed by it, his inane look carrying him towards the verge of imbecility. It was easy to see him wondering what on earth that big fuck off red button could be. It consumed him. He did have a slight Neanderthaloid look to him anyway, but now it was pronounced as his brow furrowed, shaped as if ready to headbutt a dinosaur. He tried to look away but his obsession forced his glazed expression towards the big fuck off red button. He slowly averted his gaze again, only to whip a stern glare back to it. Before anyone could warn him not to touch the big fuck off red button that, by now, had hypnotised him, he pressed it. He fucking pressed the big fuck off red button. Even more amazing is that he looked surprised when the emergency klaxons sounded and matelots started to run around like headless chickens to try and save the poor sod that had apparently fallen overboard.

Monty was castigated by many, but supported, no matter what, by the bootnecks on board. What many didn't know, behind Monty's facade of stupidity, was his highly successful property portfolio and his expedited progress towards financial freedom. To his credit, he told very few people and never used it as defence when being accused of being an idiot, which was often.

Time at sea was hectic due to boardings, MGD(V) duties, weapon training the ship's company and looking after the small arms store. Shore visits were a welcome relief for everyone and as I stood staring over the 4.5" gun towards the grandeur of Mount Etna I couldn't wait to see the delights of Catania, the Sicilian town that the volcano glared over.

I'd love to be a travel writer. If I was asked to write an article on Catania it would be very short. It would read, "Never, ever go to Catania. It's shit."

Maybe if I visited again on my own, I may look upon the place with a different eye. The buzz of city life heartens me, but in Catania port, I watched battered scooters honk incessantly through the smog at battered Fiats on pot-holed roads awash with litter. It felt a little sinister, especially as it is was here the Mafia was born and still productive. I hoped so. We needed protection from the gangs of youths that patrolled the town eager for a fight with the newly arrived British servicemen. This wouldn't have been a problem should we also be armed with knives, bottles and other implements of pain.

The town's architecture was stunning even if covered in urban soot, but we found very little in the way of a welcome, only a local football team showing any peaceful interest in our arrival.

The only redeeming feature for many of the ship's crew was that the newsstands that dotted the town sold hardcore pornography. Sat appropriately on the shelves next to 'My Little Pony' comics, eye-catching photographs showed various acts of naughtiness, usually of naked ladies with man parts stuffed inside them. The newsstands became rather a popular destination, yet I knew no one who bought 'My Little Pony' comics. I didn't know many who could read the Sicilian newspapers either, so it didn't take the skills of Sherlock Holmes to guess the most sought after genre. Some of the less virtuous matelots even binned some of their civilian clothes from the tiny mess deck lockers to make room for their stash of new adult literature.

Whether due to warnings by the police or the fact that many matelots had been chased around the city by gangs of weapon-wielding youths I don't know, but our shore leave was cancelled after only two nights. There was little disappointment and the Captain must have been happy that the crew couldn't wait to get back to sea.

We received a new Detachment Sergeant Major, Jeff - a jovial, barrel chested DL I knew from 40 Commando. He wore a permanent smile and I imagine him as a bit of a lad in his early days. He drew positives from the bootneck character, in stark contrast to the ship's hierarchy who often saw a bootneck as a bag of mental flaws. Jeff's acceptance stood him in good stead, for the lads were often in the shite, usually when venturing ashore after a few weeks at sea.

While alongside in the beautiful city of Trieste, we sat alongside the USS John F Kennedy, a monstrous 80,000 tons of super carrier that dwarfed everything around it, including HMS Cardiff, now projecting a rather pathetic tug boat look. We liaised with the on-board Navy SEAL team and shared some tips, and did some phys with them. They were big guys, the human equivalent of the ship upon which they sailed, but it helped little when challenged to the activities we were better at, such as running or snorting vodka while wearing a stupid hat.

At sea, the JFK was a totally dry ship i.e. no alcohol was served. When alongside, however, huge marquees were erected to hold leisure evenings where the crew could have a few beers before heading ashore. Obviously, with a few drunken US sailors knocking around, the US maintained a highly disciplined task force so sent out over 200 crew members just to do shore patrol. It was laughable when the Cardiff sent out a PO and four seamen to walk around the town attempting the impossible by keep the crew out of trouble. It was in one of these huge marquees where I first saw Jo out of her uniform. Stood with other Wrens in her leather jacket, hair pulled tight in a pony tail and only the slightest touch of make up, I was incredibly attracted to her. So much so, I ignored her.

Yet it seemed the attraction was mutual. On our next shore visit to Bari, we became a couple after stealing a secret kiss, seduced by the pheromone infused aroma of the disinfectant blocks of a unisex toilet. It was hardly a secret that we could keep. In fact the bootneck detachment was suddenly turning HMS Cardiff into the 'Love Boat'.

As a (non) detachment, we had no preconceptions where female sailors were concerned. The 20 or so Wrens on board HMS Cardiff were only the second batch of sea going females in the Royal Navy. Even though they were no longer officially the Women's Royal Naval Service and integrated fully within the Royal Navy, they unofficially retained their acronymic name of WRNS. The male sailors on HMS Cardiff, and no doubt elsewhere in the fleet, knew Wrens as 'turtles' because, 'once they're on their back, they're fucked.'

Their presence on board certainly polarised opinion. It would be fair to say that in the early days, many salty old sea dogs were aghast when they first boarded ship. Accusations that they couldn't do a man's job were soon dismissed as the girls proved their worth. Many of the girls I met on board were switched on, intelligent sailors. While physically they may have lagged behind the men, although there were plenty of male sailors who needed to seriously look at their waistlines, in an environment such as the operations room utilising technical skills or diligently operating radar, they were more than a match for their male counterparts.

Out of the ten Royal Marines on board, six of us were in relationships with the girls on board. This was hardly out of desperation from either side. But we found commonality with the girls. They loved a laugh, loved to take the piss out of the male sailors and had that open mindset that being in the forces amplifies. The first couple to get involved and start the ball rolling was Charlie (a girl) and Paul McGuigan. I bumped into Paul many years later in Basra, Iraq as we worked as civilians with the same company. He and Charlie were still together and had just had a son when Paul would make front page news, being one of the two Private Security guys shot and killed by a colleague, an ex-Parachute Regiment guy, while on a contract in Baghdad. To lose one's life by the enemy is hard enough, but to lose a loved one through the actions of someone who is supposedly a friend; I cannot imagine the pain of those close to Paul. Paul's murderer is now doing a life sentence in a Baghdad jail.

The bootnecks involved with the Wrens certainly did cause some good-humoured ribbing, but amongst some of the senior rates it was clear our closeness to the girls wasn't liked. It was hypocrisy of the highest level when married POs were dishing out shit claiming our relationships compromised the operational efficiency of the ship, yet as soon as the ship pulled alongside the next port they were sniffing out brothels or swarming around the available Wrens like dogs on heat.

Back on board, a rule known as the 'three foot rule' was introduced, meaning a male could not come within three feet of a female sailor unless for operational reasons. Not only was it unenforceable, as the passageways weren't even three feet across, but also highly degrading to the professionalism of the ship's crew. It was hardly likely that people would just stop and have sex while crossing each other in the galley. But enforced it was, and while certain POs would turn a blind eye to male sailors getting shitfaced or turning up pissed on a part of ship when on operational watches, they were happy to jump on anything that involved people of the opposite sex. Even the Joss - the ship's disciplinary warrant officer - took a common sense approach. The anti-bootneck POs, however, were eagle eyed if I should even talk to Jo. I can't remember how many times I was bollocked, given a move on notice, or threatened with a charge for talking to her, but if I had a penny for each time I did, I think I certainly would have been able to afford a nice pair of socks. We were grown adults in an operational arena and I was being made to feel like a naughty schoolboy, just in case I tried to peck my girlfriend behind the school bike sheds.

Our ship boardings were becoming less frequent, so most of our time was spent on the upper decks on defence watches and excess energy was building within, bootnecks chomping at the bit to get on a boarding or use any available space as an improvised exercise area.

Any sea in early summer is a beautiful place, yet having to wear anti flash hoods and gloves permanently did little for our sun tans and we looked upon, with a touch of hilarity but a dash of envy, the other naval ships we passed, such as the Italians and the Dutch, who thought it best to undertake the same operations by laying on towels, dressed in just shorts and suntan oil.

We looked forward to our time ashore where we visited ports such as Bari - again due to an engine failure, Cagliari, where the RM (non) detachment managed to get away into the mountains for some light 'training' - in truth a yomp into sanity where we could detach ourselves from the ship's regime. We also visited Naples, which has the character of a crack addict - dirty and violent, yet is surrounded by the beauty of the Amalfi Coast and the Island of Capri where Jo and I spent a wonderfully expensive few days on R'N'R. We took a jeep from Chania in Crete and drove through the island's rugged charm. The trip to Corfu would have been even better if I hadn't returned to Ipsos with the lads, got naked and ran up and down the beach and into the sea with a Wren from the ship that wasn't Jo, which made it slightly less funny for her.

Whilst the ship's officers were determined to change our uniform from DPM camouflage to the blue Royal Navy 'eights', we became 'Fortress Royal' defending our right to wear our uniform, not that of the ship's crew. Their reasoning was varied and could be justified, yet so too could the counter arguments, so we remained proudly in green. This stand off also became the subject of our tour t-shirt. Unsurprisingly, the RM tour t-shirt didn't go down too well with the Skipper, who banned it forthwith. It wasn't that it had a pair of hairy testicles hanging out of underwear with the word 'Pants tour', on the breast; or even the motto 'Real Men Don't Wear Eights' around a picture of an armed Bootneck on the back. No, the Captain was livid due to the phrase 'RM Detachment.'

It would be fair to say we were excited as we sailed homeward bound into Portsmouth Harbour. Some Navy families did the traditional thing and stood at the roundel with banners welcoming home their loved ones. I found this rather touching. It was a real homecoming, something I'd not yet experienced in the Corps, and felt a welling of pride as I stood on the forecastle in my best bib and tucker. It would also be my first chance to meet Jo's family, so ideally attired.

With the operational tour behind us, all the ship had to look forward to was a few months in dry dock while it undertook a large maintenance programme, leaving us all a little frustrated as to how best to spend our time. We went off doing military stuff, did plenty of 'phys' and went ashore rather more frequently than was healthy. The only thing I had planned prior to Christmas was a small matter of buying a house in Taunton with the money I'd saved while away, and an operation on a part of my body my gran would describe as 'down below'.

*'Mummy, when I grow up I want to be
a Royal Marines Commando.
We'll decide son, you can't do both.'*

~ *Internet meme*

BUYING A HOUSE was rather a large step. Rather than seeing my future through the bottom of a pint glass, I was determined to secure some sort of long-term commitment. As Jo and I seemed to be pretty smitten with each other, buying a house seemed a logical, if not rather hurried, decision. With the ship fresh from dry dock, a trip to Gibraltar to give her a good run out coincided with me having to sign legal papers in Taunton. I was kindly given permission by my Divisional Officer to fly down at my own expense to meet the ship in Gibraltar. This was a perfect solution, one that raised eyebrows in the wardroom to those officers who thought it beyond a mere corporal to make his own way and certainly not acceptable for a non commissioned rank to 'do his own thing'.

As promised, I met the ship with the house con-tracts signed, and the keys to be collected on 23rd December. After only a few minutes after setting foot on board, one of the officers cornered me to let me know how disgusted he was that I'd been allowed to come down under my own steam.

'The ship is about teamwork. If you think the stunt you pulled is acceptable, I think you'll find you're wrong. Seeing as though you like the limelight, be aware I've got my eye on you,' he stated, his chinless head wobbling as a result of having no backbone.

As far as stunts go, mine was rather meek - akin to unicycling in a purple velour poncho or crossing the road in rather heavy traffic - so I didn't really see how I had been out of line yet, but as we had clashed many times previously on differing opinions he thought I was undeserving of having, we remained guarded behind simmering enmity. Here was a man who lived in a village that bore his surname, in a mansion that bore his surname, and with a private education that totally removed him from the realities of those not fortunate enough to have been born with a silver spoon.

'Oh, I am Duty Officer tonight and have decided you can be my duty Quartermaster,' he said with the glib smile that can only be given when the struggles of the populace are only casually noted in the broadsheets of a private club.

The role of a ship's Quartermaster (QM) while alongside is basically to be an upper deck sentry, man the gangway and ensure crew members sign in and out when they leave and board the ship. One of the perks of being a QM is that returning sailors often buy them scran from ashore. I'd already shared a burger with my boatswain's mate, Lee, one of the new Marines on board. Up the gangway came big Sid, a larger than life matelot who always wore a smile on his face especially when he had beer, fags and a kebab in front of him - sometimes all at once. He was rather a little worse for wear when he staggered up the gangway and I thanked him when he passed me a package of paper inside of which was a large portion of chips and seafood cole-slaw. Lee and I ate the lot and even turned down the next offering of food as we were now contently stuffed.

'Cheers for last night, Sid,' I said the next morning.

'For what?' he answered through bleary eyes.

'For the scran you gave us when you came back from ashore.'

His face wore a mask of horror. 'That wasn't scran for you to eat.'

'What do you mean?'

'I gave you that to put in the bin. I'd just been to the chippy and had a crabstick eating contest. I managed to eat 17 of the fuckers walking back but then puked 'em all into my chips.'

Word quickly spread that Lee and I were quite happy to eat puke, which again turned into a drama for the officer with the mansion named after his family. He launched into me yet again, 'Is this true?' he asked while on the bridge.

'Unfortunately, yes, Sir.'

He then went into a rather unnecessary tirade accusing me of being 'not fit to wear the uniform of Her Majesty's Forces,' calling me amongst other things 'an animal' 'a disgrace,' and mystifyingly 'a Chernobyl orphan', for which I found genuinely hilarious, that only inflamed his ever increasing anger.

Despite trying to ignore his elocutionary perfect tirades, I was wondering whether being on ship was worth the grief. The sanctimony of Naval hierarchy neuters even the most intellectual debate, the irony being it debases discourse to perfunctory statements upon which the highest rank is deemed correct. Of course, 'losing' in such dialogue with him didn't matter when one could secretly put boot polish around his binocular rims.

Getting back to Pompey I was now preparing for my operation. With the operation at RNH Haslar the following day, I was wary of doing anything silly on the preceding Sunday night. Wary yes, but with Jo on duty, a Sunday evening sat on the windowless mess deck listening to blokes farting; the bright lights of HMS Nelson and the infamous 'Nelson Bop' only a stagger back, the temptation was too much. I dragged out Ant, who promised to keep me relatively sober. As my operation was at 1600 the following day it was only fair I partake in a couple of shandies. What harm could it do?

My previous jaunt to the Nelson Bop was on my 19th birthday while on my second stint with the Corps boxing squad. I do recall celebrating this most irrelevant of birthdays naked on the dance floor with the neck of an empty bottle of Newcastle Brown Ale wedged up my anus - and yes it was my birthday party piece - so it was of no surprise that this particular visit would also end up rather messy. Stumbling out of the bop at midnight blaming Ant for dragging me out and making me drink so much, we headed for the rear 'Anchor Gate' - the conduit between the naval camp and the dockyard. No gate now existed so walking through the archway was far too easy. It was a far better plan to climb over the 6 metre high wall that surrounded it. The plan was so idiotic that having taken off my shoes to climb over, I'd have to pass back through the gate to retrieve them before walking through as per normal. With my adept climbing skills still at hand, I reached the top only to be shouted at by a deep voice from below. I cannot recall the exact words but it did mention:

a) I was an idiot

b) I was to get down immediately

I did politely request to just climb over the top to at least conquer the wall, after all there is little accomplishment in turning back so near the capped-stone peak. Who'd have heard of Sir Edmund Hillary if Sherpa Tenzing had asked him to turn around before the summit as reaching the top was a stupid thing to do? I promised I'd meet the voice in the dark when I scaled back down the other side. Yet, despite being so near to mountaineering glory, my request was impolitely refused. So for the sake of humanity, I clawed my way back down. The voice had a film noire face; an evil shadow cast half way across it from the indirect lights of the perimeter fencing. He asked for my ID and checked it. He didn't know me from Adam and I was rather put out when he said, 'Ah a bootneck, typical.'

It was far from the truth, I was nowhere near typical. Typical would have been to walk through the Anchor Gate. He started to drone on about something or other, I can't recall, I lost interest after the first few sentences, before stating I had to report to the gatehouse.

The conversation deteriorated to me childishly calling him a 'fucking matelot wanker' - hardly Thomas Hardy, yet did promote the notion that alcohol is the mouthpiece of truth. Only at this point did he inform me he was a Regulating Petty Officer and that I was now really in the shit. Realising that these 'regi' guys were more cut out for jobs as future traffic wardens than social workers, common sense overtook any sense of self righteous indignation so decided not to inflame the situation further by ceding to his order to follow him back to the gatehouse. En route, the Regulating PO grabbed my shoulder and pushed me forward. Now, as anyone would know, I hardly walk in a slovenly fashion so speed cannot have been his reason. As well as being rather unnecessary, my main concern was his big right hand, one that had probably been clasped firmly around his sweaty manhood not more than 10 minutes before, creasing my pretty classy jacket. Pulling away led to something else I hardly desired. Where the other four regulating staff came from I don't know, but within seconds I had four rather hefty 'regies' wrestling me to the floor. Ant joined in, but as I was now face down being choked I questioned what positive impact he was having on the proceedings. My circumstances weren't pleasant. Evidently not aware of positional asphyxiation they positioned me so that I was now face down with my head a nice cushion for someone's knee and my torso a park bench for someone who obviously enjoyed the consumption of desserts. My arms were also in an atypical position, wrenched unceremoniously behind me, and my wrists had suddenly been adorned with the bling of biting handcuffs. I recognised a heavy voice.

'Still a matelot wanker am I?'

I did repeat that my opinion hadn't yet changed, adding, 'You should get a part as a nasty man on *Taggart*.'

Why, I don't know, but he still didn't find me funny.

Yanked to my feet, I saw Ant in a similar pose with three Regulating POs on him. I had four so gave him shit through a bleeding mouth about his weakness in only three people required to hold him down. 'I,' I claimed; 'am 33.33% harder,' repeating the recurring '3' an unnecessarily annoying number of times straight into the face of the sweaty 'regi' PO.

With us told to face the wall of the gatehouse like a couple of school miscreants, we waited for the Duty Officer to come and see us. To pass away his feeling of inferiority, one 'regi' who seemed to be a stunt double for the bloke in the *'Mr Muscle'* adverts sidled up behind me and wrenched the handcuffs up my back tearing on my shoulders.

'Not so hard now, are ya bootneck?'

If I wasn't shackled I would have gouged out his eyes with a spork.

Unfortunately I was, so settled for second best, 'I'm gonna fuck you up when I get released, you weasely little shitcunt.' Eloquent verbosity had never come so naturally.

'Release? I'll have left the Navy by the time you get out,' he replied glibly.

The way I was going, he was probably correct.

The Duty Officer, tired and irritated at being peeled out of his bed to speak to two slightly merry fuckwit bootnecks was in no mood for frivolities.

'Names?' His tired intonation suggested he wanted this over with quickly so he could return to his pit.

'Sorry Sir, I cannot answer that question,' said Ant giggling to me like he was a schoolgirl with a lolly.

It seemed the Duty Officer's expedited return to slumber was a rather quixotic fantasy.

'I haven't got time for your games. Name?' he said with a heavy sigh. It was hardly barbaric questioning.

'Sorry Sir, I cannot answer that question.'

And so it continued. Well, what could he do? Make us pregnant? We continued past his loss of temper, until even we got bored of our own self-indulgence.

Once names were ascertained (he didn't believe Ant when he did say his name was Kuntos McTavish), he called for the Doctor who was as impressed as the Duty Officer at being dragged from slumber just to confirm we were sufficiently inebriated to justify a charge of drunkenness.

Passing the medical test to confirm our state of intoxication (I love passing tests), we were carted away to Portsmouth Detention Quarters, the Victorian naval prison that remained staunchly Victorian in its regime.

Having been removed of my shoelaces - I was hardly a suicide risk I was being sent down for a night, not a 30 year stretch, besides my shoes only had four holes so my laces were only long enough to choke a snake - I was thrown into a cell and dejectedly laid down on the wooden bed. No sorry, I can't call it a bed. It was just a block of wood. Wood, I think, that was in the top percentile of hardness in the Janka wood hardness scale. Only a few rusty nails embedded would have made it more uncomfortable. Needless to say, the night before going in for an operation where I'd been told to get a good night's rest, I was having quite the opposite. I was frigging freezing, but would not allow myself to ask for a blanket, that would show weakness to the matelots. I hadn't thought I was particularly drunk, but spent the majority of the night showing my weakness by puking purple bile into the shiny spittoon that was my sole companion for the evening. I was confused - I hadn't even eaten beetroot.

I was awoken by stripy sunlight. My head was in a state of turmoil, as was my pea-sized stomach. To convalesce me was a huge old colour sergeant boot-neck, who ordered me up, empty and clean out the vomit filled spittoon and follow him to the reception desk that I recognised from about five hours earlier. I looked at Ant who I hoped looked worse than me. We were handed back our laces and told to sort ourselves out. Good idea.

We walked back to the ship thankful for the tonic of cold fresh sea air. As we approached the ship I saw the familiar outline of Jo on the flight deck. She was on gangway duty. Usually she smiled when she saw me. This time she shot me an obsidian look. A happy bunny she was not.

'Are you mental? I've been worried sick. What are we going to do about the house?'

I was confused.

'They're going to lock you up for ages.'

I know I was hungover, but my cognition was certainly struggling to take in her words.

I then vomited on the flight deck. It wasn't the most auspicious entrance I'd ever made.

My mind as clear as my stomach, it then transpired that Jo was at the receiving end of a chain of Chinese whispers. Far from just being a couple of climbing bums sent to DQs for the night to sober up, we were desperados who had beaten up a few Regulating POs, been arrested, then sent to DQs. After disarming the guards (not that they were armed), we'd escaped from custody and were now rampaging around the dockyard armed and dangerous. I was impressed; in reality if DQs had been one big paper bag I'd have struggled to punch my way out.

I was still feeling a little hungover and rather nauseous from fasting in the way a bulimic does, when Jo, relieved from the worry of any future relationship separated by iron bars, dropped me off at the Royal Naval Hospital in Haslar.

Whisked into theatre, the operation wouldn't be a lengthy or even complicated process. I was being circumcised. My foreskin had become a liability and needed binning. I could, at this moment, pretend that it was due to me being hung like a prize-winning horse, but I cannot lie. I just had a rather perishable foreskin, a sort of weathered washer like you find on a cheap tap.

Asking if I could keep it for a souvenir just before going under, I awoke feeling like shit. Whether the hangover combined with post anaesthesia grogginess took hold, but I wasn't really feeling like I could do much else but feel sorry for myself. I had little pain from the area of surgery and my first peek was of a cock that looked like the original blind cobbler's thumb. It was wrapped heavily in bandages and I dared not have a look at what lurked within. My request to keep my detached foreskin hadn't been taken up so my plan to give big Sid the matelot a revenge calamari ring never materialised.

I was awoken again at 2am. Not for prescribed drugs, or by the onset of pain, but by the ward nurse who informed me I had a phone call. Limping forlornly to the ward phone I was confused as to what emergency I could rectify with a traumatised cock and my arse hanging out of the back of a poorly tied hospital gown. There was no emergency. It was Jo. Pissed as a parrot in a phone box cuddling a 'Messy Burger'. She was missing me, and wanted me to come and take her back to ship. In my state it wasn't going to happen. Trying to verbally arrest her grip from the phone, she kept burbling that I no longer loved her, and to prove my love I was to come and get her. I was annoyed; not at being awoken by a drunken fool burbling their undying love from the confines of a piss stained phone box; but by the fact that in our relationship, that was my job.

I can only assume the RN surgeon who conducted the operation must have gone to the same medical school as Josef Mengele. He must have left his surgical utensils at home so improvised using a rusty carving knife and an orange peeler. He had managed to transform my lovely smooth and rather charming penis into Frankenstein's cock. However, I proudly communicated and showed anyone who would listen/look that the average number of stiches for a circumcision was seven. I had fifteen. I then realised the average age of a person undergoing circumcision would only be about two days.

Due to the discomfort of having a penis that was permanently twice its normal size plus barbed wired-like cat gut stitches grating anything close, I walked around with my hands inside my trousers until someone noted I looked like a seedy pervert.

If I hadn't enough trouble with my own penis, I had those dicks from HMS Nelson on my back through statements that they'd written in support of my upcoming charge sheet. It didn't read well. I was being charged with being drunk and disorderly, insubordination, resisting arrest, assaulting a Regulating Officer, and the recent assassination of Melchior Ndadaye, the President of Burundi. The first charge I couldn't deny, but the other charges were serious enough to lose my rank and see my career take a huge step backwards. When in a commando unit we worked under Army regulations. While stringent compared to civilian law, I would have to kill the CO's children to lose my rank. However, on ship we came under the far more draconian Naval Regulations where I could lose my rank for something as little as wearing my beret at a jaunty angle.

I was now literally shitting my DPM kecks. From a stupid prank I was now in dire trouble. The Joss man, who I regarded with the utmost respect, showed me the regi POs' statements and at least I could see some amusement. I was pleasantly surprised to be heralded as the new Jean Claude Van Damme. They were clearly works of fiction as in the circumstances, I was hardly able to punch out a recalcitrant budgie, let alone take out a group of men with the combined weight of the Titanic. The Joss also recognised these statements had more holes than a tramp's undercrackers. Within 48 hours the charges had been downgraded to drunk and disorderly and insubordination.

That evening my charge sheet was close to getting an addendum. As part of the rehabilitation of my traumatised bits, I had been told to let my penis 'air' after washing it - carefully I may add - evidently after showering was the best time. Shower time on ship is always a mad rush, everyone trying to get ready and the heads then scrubbed prior to evening rounds. I'd just dried off after my shower, and stood in my gulch awaiting rounds wearing just a towel that I held loosely due to my cock being too painful for it to be tightly wrapped. Being donned in a towel wasn't necessarily a problem when alongside for evening rounds but having to hold my towel with one hand meant I couldn't properly stand to attention. I did my best, with heels together and one arm down my side when the inspecting officer came down the mess.

'Corporal Time, why are you not stood properly to attention?'

How do you say politely that you have a sore cock? I could have just said, "Cos I've got a sore cock,' but thought that would sound rather facetious. I settled with, 'I've just had a rather delicate operation, Sir.'

'And?'

'I can't wrap my towel tight, so I have to hold it.'

'I can't see how that works. Stand properly please.'

So I did. Lo and behold, due to the laws of gravity that seemingly he wasn't aware of, down fell my towel. There I was, starkers, stood to attention with my stitched up black and blue dick dangling forlornly before his eyes. I don't know whether he'd been abused at public school but the sight of my bare, swollen penis sent him apoplectic, and he warned me that along with all my other charges that had mounted up I was to be charged with indecent exposure. I could have laughed this off, however, while I could accept being a drunken insubordinate idiot, a charge of indecent exposure could put me on a sex offender's register.

The following morning, I saw the Joss man and explained my case. He understood and told me not to worry. But I did. It seemed I could do no right with the hierarchy of the ship and it was only time before my feistiness would see me doing something I'd later regret. For the first time in my career I was saliently unhappy. When such an emotion becomes overriding, a change of environment can be the first step to recovery and I feared being on ship would just give strength to my dark twin. I asked Jeff the DSM to see if I could get an early draft from the ship. Using my network of spies, I heard that a corporal from 40 Commando was looking to get away from Taunton as quickly as possible due to marital problems so was happy to swap drafts with me. I would be free from this hell hole straight after Christmas leave. All there was to do before I left was to put up Christmas decorations and be charged.

Thankfully, it was a private affair at the Captain's table, yet I was doomed as I walked into his office. I could already see my punishment written out while he listened intently to the Executive Officer's, acting as prosecuting officer, account of the night in question. His rather posh accent was quite comical when he recounted my turn of phrase 'you matelot wanker', bringing a snigger from Jeff who, in return, got a rather pathetic bollocking from the Captain. The Regulating PO who I'd first met while he attempted a right arm gaston up a wall, stepped in and lied like a politician's expense sheet. As the bastion of Naval Regulations, one would have thought he would have upheld the virtues of integrity and honour when providing the facts upon which the case rested. It would therefore be easy to see why the ship's Captain would believe his word over some scruffy arsed bootneck with an infamous short temper and a mouth that saw deference of position an option. I tried to interrupt, suggesting his untruths, but was in turn told to shut up. Becoming crimson with frustration, like an incoherent toddler requesting his favourite TV channel, I stood there listening to more bullshit hoping parity would be restored once my Divisional Officer stood to give my defence. His well-balanced mitigation may have been some use if the Captain had listened to him but, instead, started chatting to the Jimmy, totally ignoring my DO. I stared in disbelief. This was the highest form of ignorance and at least contravened the basic fundamentals of natural law. No wonder Queen Victoria named Naval Officers 'pigs'.

I was a beaten man. When it came to my turn to give evidence I offered little. There was no point. It was a kangaroo court. I was the guilty bastard even before I'd walked in.

The Captain summed up by saying I was a security risk. Not being the time to argue, but I did think that if I was indeed a terrorist, managing to already be inside the camp would surely throw suspicion onto the incompetence of the gate staff. And although the jibes of Irish intellect are well documented through various generations, I would suggest even the most stupid of PIRA operators would try to avoid drawing attention by not getting drunk and climbing bare footed over a wall where a perfectly adequate passage existed.

It mattered not. The Captain fined me £400. He understood I was moving into a new house so allowed this fine to be paid in two instalments, so I sarcastically thanked him for his compassion. He was not done yet though. He also gave me ten days '10's' which meant I would be confined to ship and ordered to parade at silly times and be looked upon with disdain by the children who had the audacity to hold a commission. The ten day curfew would mean I would be released on Christmas Eve. Jo would move into our new house on her own. *Happy fucking Christmas, Captain.*

It was a rather anticlimactic ending to my time on ship, a draft I'd volunteered for in the haste of travel urges but regretted in the leisure through my intrinsic resistance to naval authority. Yet I'd met the woman of my dreams and while I'd been told, 'A pat on the back from a matelot is just a recce for the knife,' I was honoured to have served alongside the majority of junior rate matelots of either sex and a minority of the seniors. It had not been a wasted year by any stretch and one that certainly reinforced the belief that absence makes the heart grow fonder.

***

After a year hiatus, my time to return to commando operations was upon me. I was determined not to let the past year stain my opinion of military life, there needed to be troughs to appreciate the highs and I felt peaked to get back to 40 Commando. I found pleasure through just the simple act of walking through the gate of Norton Manor Camp with a couple of nervous young marines fresh from basic training. One looked older than me yet his perceptions were of a man far younger. There was no stapling of chits to heads this time as there was on my first day. I wasn't a naïve 17 year old, but a 23 year old with a lifetime of experiences with an urge to find more. The Royal Marines was my racing car to adventure, yet my ambulance when consumed with darkness. I was now wise to what tours, courses and personalities I could look forward to getting involved with and I was as keen to succeed as I was on my first day. I no longer felt nerves but confidence, and the pride of being a bootneck still pulsed through my veins. It occurred to me that my career had come full circle.

I sauntered into the guardroom. The Guard Commander looked at me. 'Bleeding hell Mark, I haven't seen you for years, you raving hornblower. Still resident pole dancer at the Blue Oyster Club?'

All I could do was laugh. I was home.

Epilogue

A dark cloud seemed to hang over my beloved 40 Commando. Like returning to Narnia, things had changed. No longer the 'Sunshine Commando', '40' had become the 'Dark, Wet and Miserable Commando'. Morale was lower than a snake's belly that had been locked into a stone box and thrown into the deepest depths of the ocean. Seemingly it was down to one man - the Commanding Officer. Known as 'The Moose', the overt nom de plume of an invidious bully, reports filtered back from Belfast of how he ordered a young lieutenant to stand with his heels together on the Falls Road to the hilarity of the local PIRA supporters, and of how he threatened every newly commissioned officer with enforced resignation should they stuff up. Reports further suggested he made everyone wear issued clothing rather than their own, as was usual in a commando units, and how he turned lively marines, who previously had invincible enthusiasm to take on the world, into jaded, powerless serfs.

Men were submitting requests to leave the Royal Marines quicker than the clerks could handle. In short (although he wasn't, he was well over six feet tall), he was described as a domineering meglomanic despot with only his self-absorbed interests at heart. Also despised, with whom the CO stood shoulder to shoulder, was his sycophantic myrmidon - the RSM (or rather shoulder to head as the RSM was a short arse). RSMs nearly always commanded respect; however, his office seemed to be an appointment of begrudging deference.

Initially, I'd wanted to join the unit in Belfast, but with no places available, it appeared rear party was the best place to be. It did allow me to sort out the house and do up any odds and sods that was necessary to convert it from a homage to 1970's kitsch to something that appeared slightly stylish.

I'd spent the day battling with a steamer trying to avoid scolding my face while removing the woodchip wallpaper that had seemingly been super glued to lime plaster. The phone rang, so looking like a half finished Blue Peter model, I walked precariously down the paper ridden stairs to pick up the receiver. The operator asked if I'd accept a reverse call. Between spitting out bits of woodchip, I accepted. It was Jo. She was in tears.

'What's wrong?' I asked.

Her reply was two words long, just two simple words with enough explosive content to turn my world on its head.

'I'm pregnant.'

She was calling from Glasgow Airport, wearing her uniform, with travel warrants her only possessions. She didn't even have her purse.

Two days previously, she'd been sailing somewhere off the coast of western Scotland as part of a live firing exercise. She wasn't feeling 100% but, by her own admission, just wanted to 'pull a sickie' and after being prodded and poked by Daisy the ship's medic, was told to get some bed rest. She reported as instructed the next day to the Doctor on board, who also felt her stomach. He asked her if she had a dry metallic taste in her mouth. Jo thought this an odd question but yes, she did have a dryness she'd not encountered before. Within an hour, the Sea Dart live firing exercise was halted, the ship was called to flying stations and before she knew what was happening, she was being called to the flight deck and hurriedly ushered onto the Lynx helicopter then flown straight to the military base at Benbecular on the Outer Hebrides. She was panicking. They were halting a multi million pound exercise for her feigned illness.

Landing at Benbecular, the awaiting party was slightly confused. As it had been an emergency flight, they'd expected to receive a dead body, not an embarrassed female. Jo was sent straight to the medical centre, where a Doctor conducted a urine test to confirm she wasn't dead.

'Congratulations,' he said.

'So I'm not dead then?' she asked.

'Certainly not. In fact, you're the opposite. You're pregnant.'

The doctor probably expected a response other than her bawling her eyes out. With the old bulldog spirit of the military, he awkwardly told her it would be 'alright' then ushered her to the airfield where a flight would be leaving for Glasgow in the next half an hour. She was quickly given a ticket to Heathrow and a train warrant to get her down to Taunton.

Becoming a father was something I'd always wanted, but hadn't readily prepared for. Over the next few months I made it my job to ensure that any child of mine would get the best start in life. I did this by running around with my head up my arse wondering what the fuck I should do. I did the easy things, such as decorating the bedroom that would now be called a nursery, and bought as many colourful toys that one could muster. But the real tests I'd so far endured, my childhood, my time in basic training, beating off of my dark twin, the rare moments of abject fear while flying into hostile territory or free climbing crumbling masonry all paled into comparison. My biggest challenge was how I would balance my temporary job as a Royal Marines Commando with the most important occupation of the species - parenthood.

Printed in Great Britain
by Amazon